A Wild Australia Guide

DANGEROUS & DEADLY WILDLIFE

AUTHOR: **TED LEWIS**
PRINCIPAL PHOTOGRAPHER: **STEVE PARISH**

Contents

Opposite: King Brown Snake. **Right:** Funnelweb spider.

Introduction

Native killers lurk in every corner of Australia! This is the mythology of "Dangerous and Deadly Australia". Yet Australia is merely home to many (often small) animals that have well-developed defences against being caught and consumed. Surviving in every imaginable habitat, from ocean to outback, the country's "lethal" vertebrate and invertebrate species include representatives from a wide variety of families — and many have aroused a near-mythical reverence for the terrors they collectively manifest as "death by animal".

Our fascination with dangerous and deadly wildlife is evidenced by the number of books and top ten lists that deal with the subject. Not to mention our admiration for the David Attenboroughs, Harry Butlers, Crocodile Dundees and (the late) Steve Irwins of this world. Australians love nothing more than a good yarn and seem more than happy to regale overseas visitors with terrifying tales of killer sharks, giant spiders and lethal snakes. Like sporting achievements and natural and cultural attractions, Aussies take great pride in their dangerous fauna. This is the country's "dangerous and deadly" mythology at work.

In many ways, such mythology — a promotion of Australia as a land where only the tough survive — may ultimately damage our wildlife through gross misunderstanding. In the past our dangerous and deadly natives have been unjustly vilified or lampooned to the point that our public perception, as distorted as it is, became the dominant cultural paradigm. In the 1960s, tales of killer giant clams were common. So too is the long-lived perception of the Tasmanian Devil as a creature of uncontrollable ferocity — all thanks to Looney Tunes cartoons. These are mostly harmless distortions. A more detrimental case of this "culture kickback" is the Greynurse Shark, which suffered wholesale slaughter in the 1950s by powerhead-packing sea cowboys ridding the oceans of these supposed killers. More recently, the White-tailed Spider has suffered from human prejudice — its necrotic effects were widely sensationalised in the popular media with little scientific evidence to substantiate the claims.

CROC ATTACK

WEIPA

5PM LAST NIGHT

Normanby River Lakefield National Park

This is where a crocodile lunged and dragged a 60-year-old Townsville man from a canoe into the Normanby River about 5pm yesterday. His shocked wife fell from the canoe but managed to swim ashore and drive 20 minutes to a nearby ranger station to raise the alarm. Searchers last night located the canoe, but not the man.

COOKTOWN

Damon Guppy and Derek Tipper

Continued Page 2

However, time is a great transformer and we know that public attitudes can change. Creatures that were once reviled are now pin-ups for conservation. The White Shark, for example, is now protected in Australia, New Zealand, South Africa and parts of the United States. Despite scant scientific knowledge of its true behaviour and numbers, there is a feeling in the air that we must cherish what wildlife we have. To the extent in some cases, families and friends of shark attack victims are appearing on television advocating the rights of these fish to live — rather than calling for their heads. Finally, perhaps, we are beginning to recognise that we (all creatures) share one environment, one Earth.

Opposite: The minor necrotic effects of White-tailed Spider bites have branded these creatures as "flesh-eating". **Clockwise from top left:** Images to feed the imagination — "Croc" newspaper headlines; An Estuarine Crocodile; White Shark cage dive; Delight in holding a python.

Defining
Dangerous & Deadly

Just what makes an animal "deadly" or "dangerous" is largely a matter of perspective. To the average person, a cursory roll call of Australia's dangerous and deadly is certainly impressive, and media reports regularly endorse the country's standing as home to the most venomous snakes, spiders and marine invertebrates in the world. Such lounge room observation, however, only tells half the story.

Consider the serpent. The global data set for snake bite fatalities is speculative at best. While no comprehensive studies have yet been performed to evaluate human deaths caused by snake envenomation, a rough estimate from the World Health Organisation puts the number at more than 125,000 annually. Like most developed countries, Australia's contribution to these statistics is minimal. Approximately 3000 snake bites are recorded in Australia each year. Of this number, only 200–500 cases are severe enough to warrant the use of antivenom and only a tiny percentage (one or two cases) ultimately result in human mortality. This is all despite the fact that Australia is home to the top ten most venomous land snakes in the world according to LD_{50} toxicology tests*.

It is important to note that a snake's status as venomous should not automatically qualify it as dangerous. The Inland Taipan (*Oxyuranus microlepidotus*) is regularly "rated" as the world's most venomous snake — nevertheless, the species has never been positively implicated in a single human fatality. Two reasons for this are the fact that it lives in remote inland areas (places with extremely low population densities where people seldom visit) and its natural timidity. Despite its alternate common name of "Fierce Snake" the Inland Taipan, like most animals, prefers to avoid contact with humans, or *Homo sapiens* as much as possible. By contrast, the Eastern Brown Snake (*Pseudonaja textilis*), though less venomous by LD_{50} standards, is decidedly more dangerous — historically causing more than half of Australia's total snake bite deaths. The reason for this is similarly twofold — it is common and comfortable living in areas of concentrated human population and has a highly nervous disposition.

Cold statistics tell us that deadly animal attacks are a rarity in day-to-day life. Very few native animals go out of their way to cause us harm and often many creatures are dangerous only because we place ourselves in positions that allow them to be (swimming in tropical seas during the stinger season, for example). Viewed in this light, the deadly and dangerous rating applied to each species profile should be considered more a measure of potential rather than a complete assessment of an animal's lethality.

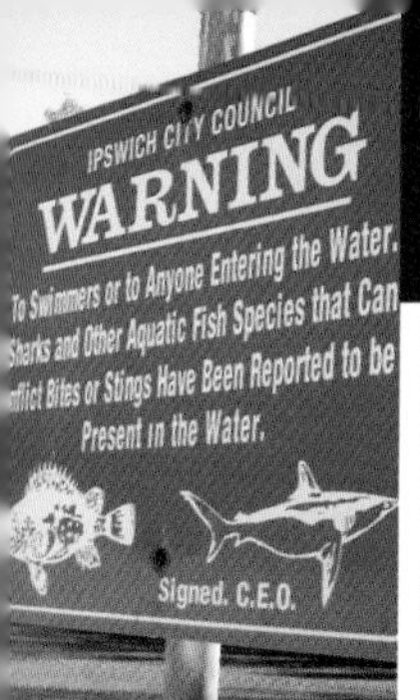

Above, clockwise from top left: Shark warning sign; Marine Stinger warning sign; Crocodile warning sign; Snake warning sign; Coastal Taipan. **Opposite:** The venom-injecting fangs of this Bird-eating Spider may be up to 1 cm long.

Animals with this label have been known to cause death or serious injury	! Very Dangerous
Animals with this label have rarely caused death but have the potential to kill	! Dangerous
Contact or interaction may result in injury, discomfort or illness	! Treat with caution
Animals with this label represent little risk to humans	Harmless

The rating system above is based on the Australian Venom Research Unit (www.arvu.org) categories. We have extended the criteria to include non-venomous but nonetheless dangerous animals.

* LD_{50}: mg/kg in saline by subcutaneous injection in mice. See page 10 for an explanation of LD_{50}.

Putting it into Perspective

Statistics and the mathematics of probability shed an illuminating light on Australia's dangerous and deadly animals and our relationships with them. The table on the opposite page demonstrates how minimal their contribution to Australia's mortality rates is, but it also tells only half the story.

A case in point is the shark. From 1791 (the time from which reliable records can be traced) up until to 2007, 191 fatal shark attacks were documented from Australian waters — an approximate rate of one attack per year. In one year alone (2003–2004), the commercial catch rate in Western Australian waters for a single species, the Bronze Whaler, was 384,464 kg of live sharks. Keep in mind that this figure does not represent sharks caught by recreational anglers, illicit shark finning operations, or government-controlled shark net and drum line programs. The Australian Shark Attack File credits this species with just three fatal attacks in local waters and yet this figure conservatively equates to the slaughter of 4000 Bronze Whalers in a single Australian State. While you ponder these figures, also consider this animal's slow maturation, low reproductive rate and ecological value as an apex predator and it becomes clear the environmental scales are not balanced.

Anyone who doubts the probability of being injured by a big fish should compare Australia's shark attack statistics with the national road toll. Records of motor vehicle accidents have been kept since 1925 and in that time more than 170,000 deaths have occurred on Australia's roads. Together with suicide, road accidents are the leading cause of external death in Australia and claim thousands of lives every year. The simple truth is this — most people will never encounter the majority of animals listed in this book outside a zoo or aquarium. You are far more likely to be attacked by someone's pet dog or run over by the family car than you are of being attacked and killed by any of these creatures.

Left: Depletion of food stocks through commercial and recreational fishing as well as spearfishing and finning operations have decimated populations of sharks.

NUMBER OF HUMAN FATALITIES

OVER AN 80 YEAR PERIOD

	Common Name	Scientific Name	Australia	Worldwide
Mammals	Dingo	*Canis lupus dingo*	2	N/A
	Asian Water Buffalo	*Bubalus bubalis*	1	N/A
Birds	Southern Cassowary	*Casuarius casuarius*	1	N/A
Reptiles	Estuarine Crocodile	*Crocodylus porosus*	17	N/A
	Coastal Taipan	*Oxyuranus scutellatus*	6	N/A
	Common Death Adder	*Acanthophis antarcticus*	5	N/A
	Tiger Snake	*Notechis scutatus*	18	N/A
	Chappell Island Tiger Snake	*Notechis scutatus serventyi*	2	N/A
	Eastern Brown Snake	*Pseudonaja textilis*	19	N/A
	King Brown Snake	*Pseudechis australis*	1	N/A
	Rough-scaled Snake	*Tropidechis carinatus*	1	N/A
	Eastern Small-eyed Snake	*Cryptophis nigrescens*	1	N/A
	Copperhead	*Austrelaps superbus*	1	N/A
	Sea snakes	Various spp.	0	150
Amphibians			0	N/A
Insects	Bees	*Hymenoptera* spp.	38	N/A
	Wasps	*Hymenoptera* spp.	7	N/A
	Bull ants	*Myrmecia* spp.	6	N/A
Arachnids	Sydney Funnelweb Spider	*Atrax robustus*	13	N/A
	Red-back Spider	*Latrodectus hasselti*	14	N/A
	Paralysis Tick	*Ixodes holocyclus*	20	N/A
Fish	White Shark	*Carcharodon carcharias*	40	232
	Tiger Shark	*Galeocerdo cuvier*	23	86
	Bull Shark	*Carcharhinus leucas*	10	75
	Hammerhead sharks	*Sphyrna* spp.	0	16
	Bronze Whaler Shark	*Carcharhinus brachyurus*	3	15
	Stonefish	*Synanceia* spp.	1	4
	Stingrays	Various spp.	3	17
Jellyfish	Box Jelly	*Chironex fleckeri*	67	N/A
	Irukandji	*Carukia barnesi*	2	N/A
Other marine animals				
	Cone shells	*Conus* spp.	1	15
	Blue-ringed Octopus	*Hapalochlaena maculosa*	2	1

Venom

Australian venoms and biotoxins are some of the most potent on Earth and serve a variety of purposes from killing prey to aiding digestion. In general, the chemistry of venom is highly complex. Although specific organisations, such as the Australian Venom Research Unit, undertake detailed studies of the composition and effects of venom, our understanding of native toxins remains patchy.

The reason for this is that many venomous species resist scientific investigation. The Irukandji (*Carukia barnesi*) is a tiny sea jelly that is almost invisible, and obtaining enough of its venom for study is difficult. Other animals, such as spiders and stinging fish, are extremely tricky creatures to "milk". Even if enough venom is extracted, testing its effects in a laboratory poses its own problems. Often test animals will not have the same susceptibility to venom that humans have. A good example of this is funnelweb atraxotoxin, which is deadly to primates but seems to have little effect on other mammals.

Some of the main components of venom are:

- **Neurotoxins** — attack the nervous system and cause paralysis.
- **Myotoxins** — break down muscle cells.
- **Haemotoxins** — affect the red blood cells.
- **Nephrotoxins** — affect the kidneys and cause renal failure.
- **Cardiotoxins** — affect the heart.
- **Necrotoxins** — break down skin and muscle tissues.
- **Procoagulants** — prevent blood clotting and cause severe bleeding.

LD_{50}

Many people ask the question: "What is the world's most venomous snake?" To answer this question requires the measurement and comparison of different snakes' toxic capabilities. One method of achieving this is known as the median lethal dose, or "LD_{50}" in toxicology circles. LD_{50} represents the lethal dose required to kill 50% of a test population (experimental mice, for example) and gives a reasonable, though by no means complete, indication of a snake's lethal "potential". According to this scale, the five most venomous snakes in the world are listed in the table below:

SNAKE SPECIES	LD_{50}	DISTRIBUTION
Inland Taipan (*Oxyuranus microlepidotus*)	0.025	Australia
Eastern Brown Snake (*Pseudonaja textilis*)	0.053	Australia
Coastal Taipan (*Oxyuranus scutellatus*)	0.099	Australia
Common Tiger Snake (*Notechis scutatus*)	0.118	Australia
Black Tiger Snake (*Notechis ater niger*)	0.131	Australia

Information from Australian Venom Research Unit http://www.avru.org/general/general_mostvenom.html

Antivenom

With so many venomous creatures scattered throughout the country, it is not surprising that Australia is a world leader in antivenom research and development. Antivenoms are purified antibodies that are injected as a serum into the bloodstream of people who are bitten or stung by venomous animals. These antibodies bind themselves to venom molecules and neutralise their effects. The development of antivenom has spared many lives from bites and stings that once proved fatal.

The first antivenom in Australia was developed in the New South Wales Health Department Laboratory by Frank Tidswell in 1898 and the method he employed is still used to this day. The antibodies contained in antivenom are made by animals — most typically horses. When a horse is inoculated with a very small, non-lethal dose of venom, its immune system creates natural antibodies to combat the venom's effects. These antibodies can then be extracted from the bloodstream and purified for clinical use on humans.

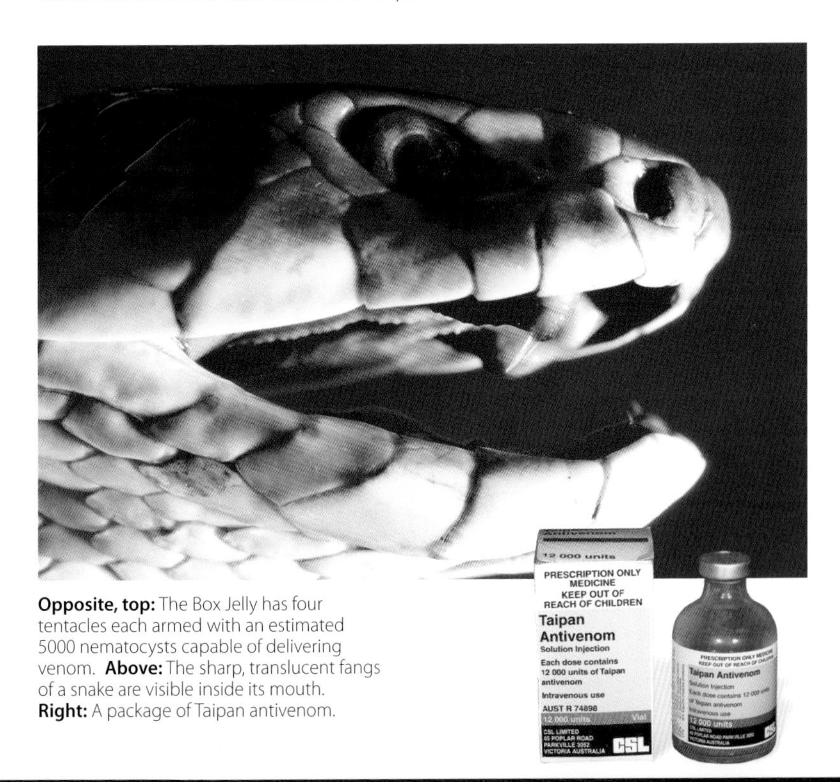

Opposite, top: The Box Jelly has four tentacles each armed with an estimated 5000 nematocysts capable of delivering venom. **Above:** The sharp, translucent fangs of a snake are visible inside its mouth. **Right:** A package of Taipan antivenom.

Reptiles

Australia is home to some 840 reptile species, contained within sixteen families, and its long isolation from the rest of the world has allowed some truly unique creatures to evolve. Most of the continent is a hot and harsh place, but it has been particularly kind to these cold-blooded creatures. While reptiles inhabit most native environments, both on land and in water, many are especially well adapted to those vast, seemingly lifeless, tracts of the country's interior.

Australia's dangerous and deadly reptiles can be divided into three main groups — crocodilians, front-fanged terrestrial snakes and sea snakes. While there is no doubting the toxicity of Australia's sea snakes, the country is most famous for its terrestrial elapids. Front-fanged land snakes are our dominant snake group and nowhere else in the world do they survive with such diversity. About 25 species are capable of killing humans with their venom, but all prefer to retreat or camouflage themselves at the slightest hint of human contact. Among them are the taipans, death adders, brown snakes and black snakes — beautiful and widely misunderstood reptiles that strike some people with fear at the very mention of their names.

Native crocodilians are represented by just two species — the Estuarine Crocodile (*Crocodylus porosus*) and Freshwater Crocodile (*C. johnstoni*). Isolated incidents of people being attacked by the Freshwater Crocodile are probably territorial responses or cases of mishandling or mistaken identity. This species is built for fish-eating and, when left alone, does not pose a significant threat to people. The Estuarine Crocodile, by contrast, is in a different league altogether. One of the largest and most advanced reptiles on Earth, the "saltie" commands widespread respect as a predator — taking on, and usually defeating, any animal that dares enter its dominion. The best way to prevent attack is to avoid swimming in areas that crocodiles inhabit.

Top: Small-eyed Snake (*Cryptophis nigrescens*). **Right:** The highly venomous Banded Sea Krait (*Laticauda colubrina*). **Opposite:** Estuarine Crocodile or Saltwater Crocodile as commonly referred.

Estuarine Crocodile *Crocodylus porosus*

Crocodilians represent one of the oldest unchanged lineages in the animal kingdom. Such evolutionary constancy is the result of predatory adaptations perfected around 100 million years ago and their body plan continues to remain unchanged into the 21st century. The giant of the order, the Estuarine (or Saltwater) Crocodile, thrives in a range of aquatic and terrestrial habitats across tropical Australia. While attacks on humans are rare, this species is rightly feared as one of the country's, and indeed the world's, most lethal carnivores.

DESCRIPTION: These large, sluggish-looking reptiles have tough "armoured" skin. Bony plates (osteoderms) run in parallel ridges along the back; according to some, these help "channel" water while swimming, thus reducing visible wake. The muted olive or slate-grey dorsal skin incorporates yellow or dark patches. The limbs are short and stout, the back feet are webbed, and the long, muscular tail is flattened on both sides. The jaws are armed with 64–66 cone-shaped teeth. Fully grown adult males in excess of 6 m, and weighing more than 1000 kg, have been reliably recorded.

BEHAVIOUR: This species lazes by day and lurks by night. Despite their size, Estuarine Crocodiles are one of the world's most effective practitioners of stealth. These ambush hunters keep a watchful gaze on activity above the water's surface, exposing only their nostrils and eyes above the waterline. Estuarine Crocodiles have been protected in Australia since 1971, allowing these creatures to freely occupy the apex niche of various marine and freshwater habitats across the country's tropical belt. Surviving in

AV. SIZE: 4.5 m (male)
HABITAT: Tropical waterways (tidal rivers, estuaries, swamps, floodplains and billabongs), beaches, offshore islands, cays and reefs

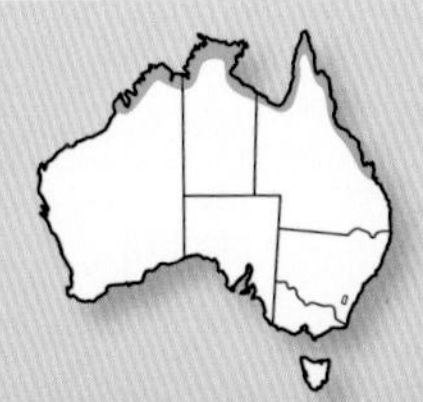

FIRST AID: May be massive tissue trauma; seek immediate medical aid
D&D RATING: Very dangerous

tidal rivers, estuaries, and landlocked billabongs of the continental mainland, Estuarine Crocodiles are also built for long-range voyaging to offshore reefs, islands and cays (where they have attacked divers).

This species' fearlessness is truly awesome — fully grown males can confidently attack and take down the largest of Australia's land animals (introduced Asian Water Buffalo, massive as they are, would prove little more than an inconvenience to a hungry and determined saltie on its own terms). Prey is typically surprised at the water's edge in an abrupt and violent assault and hauled underwater until drowned. However, other animals (such as waterbirds and fish) are taken while swimming and there are unnerving instances of wild crocodiles snatching at campers asleep in beachside tents.

One of the most alarming aspects of the Estuarine Crocodile's impressive sensory arsenal is its ability to recognise patterns of routine. Crocodilians have the most developed of all reptile brains and quickly learn to stalk prey that follow predictable pathways to water (this includes unwary humans).

DANGER TO HUMANS: There are three main reasons why Estuarine Crocodiles have been known to take humans. Firstly, crocs are highly territorial and savagely defend their patch against intruders. Secondly, breeding females are famously maternal and never hesitate to attack perceived threats to their young. Lastly, crocodiles will attack humans for food. Since 1971, there have been approximately 65 unprovoked attacks on humans in Australia. For people fortunate enough to survive an Estuarine Crocodile attack, the most common injuries inflicted are soft tissue and bone injuries. Studies have shown that most fatal attacks occur from drowning, being bitten in half or decapitation.

Taipans

The taipan's status in the pantheon of Australia's dangerous and deadly is the stuff of legend. The Coastal Taipan (*Oxyuranus scutellatus*) is the country's largest venomous snake and gained a huge degree of infamy for the number of deaths it caused in north Queensland cane fields during European settlement. Before an effective antivenom became clinically available in 1955, bites from this excitable elapid almost always proved fatal.

Two other native species are recorded — the Inland Taipan (*O. microlepidotus*) and Central Ranges Taipan (*O. temporalis*). All three yield relatively large doses of extremely toxic venom. However, it is the coastal variety of this genus that most people readily associate with the word "taipan".

All taipans are selective feeders and hunt only warm-blooded prey. They have a particular fondness for rats and mice, which explains why the Coastal Taipan is often drawn to dumps, cane paddocks, sheds and other areas around farms where rodents are likely to congregate. Taipans are superbly adapted for hunting their prey. They possess excellent vision, lightning-quick reflexes and long fangs with access to copious reserves of venom.

Taipan venom is strongly neurotoxic, rapidly attacking its victim's nervous system. The venom also contains myotoxins that break down muscles and coagulants that clot the blood. The onset of symptoms is fast and mortality without antivenom treatment is almost guaranteed. In scientific literature there is only one case of a person surviving a taipan bite without the use of specific antivenom.

Top and right: The Coastal Taipan (*Oxyuranus scutellatus*) has a liking for rats and mice.

Coastal Taipan *Oxyuranus scutellatus*

Although by nature a shy animal, Australia's "scourge of the cane fields" has a justifiable reputation as a ferocious and willing striker when provoked.

DESCRIPTION: Uniformly light to dark tan with a creamy underbelly (often freckled with red-pink spots). The head is long and narrow, and distinct from the neck. Large eyes with orange irises.

BEHAVIOUR: Coastal Taipans generally hunt during the day and employ a safety-conscious bite–release–follow strategy. This eliminates any chance of victim retaliation while their potent neurotoxin goes to work.

DANGER TO HUMANS: The Coastal Taipan's diurnal habits and taste for rodents often brings it into contact with humans. Mercifully, this handsome assassin prefers to avoid people. It should never be cornered or placed in a position where it has to defend itself.

AV. SIZE: 2.2 m

HABITAT: Dry sclerophyll forest; humid (not wet) woodland; prefers areas with patchy shade and fallen logs for cover

FIRST AID: Pressure immobilisation bandage; antivenom available; seek immediate medical attention

D&D RATING: Very dangerous

Inland Taipan *Oxyuranus microlepidotus*

The Inland Taipan is rarely encountered by humans and is considered in herpetology circles to be a much more placid reptile than its coastal kin. Nevertheless, a bite from this species should be treated as life-threatening.

DESCRIPTION: Similar body shape to *O. scutellatus* (with slender neck distinct from head). Head usually glossy black but may fade with age. Dark brown, grey brown or yellow brown above. Some scales have black edging, which imparts a flecked appearance. The undersurface is cream to yellow.

BEHAVIOUR: Like the Coastal Taipan, the Inland Taipan feeds exclusively on warm-blooded prey. This snake has adapted to a seasonally harsh and dry environment where it seeks shelter and lays eggs within cracks in the flat plains.

DANGER TO HUMANS: The Inland Taipan has bitten professional and amateur reptile handlers while in captivity, but as yet no human fatalities have been recorded.

AV. SIZE: 1.8 m

HABITAT: Arid inland areas of eastern Central Australia with sparse vegetation

FIRST AID: Pressure immobilisation bandage; antivenom available; seek immediate medical attention

D&D RATING: Very dangerous

Brown Snakes

Australia's brown snakes belong to the genus *Pseudonaja* and are responsible for more than half of the country's total snake bite deaths. Brown snakes are fast-moving, highly venomous and widespread snakes, known for their startling threat displays. When confronted, these snakes will raise the forebody off the ground in a characteristic S-shape with the mouth agape. Strikes are swift and launched repeatedly.

Brown snakes are opportunistic, diurnal hunters that feed on small mammals, frogs, birds, other reptiles and eggs. A fascinating quirk of their hunting technique is the use of constriction to subdue prey. Commonly this tactic is employed by non-venomous pythons, but brown snakes have most likely added it to their attacking repertoire as an extra safeguard against their prey escaping.

Brown snakes have adapted well to human environments. Coupled with their temperamental nature, it is therefore not surprising that they are responsible for the majority of serious snake bites in Australia. While the Eastern Brown is considered the second most venomous land snake in the world, instances of life-threatening envenomation might be a lot higher if its fangs were longer. Because brown snake fangs are short, many attacks result in so-called "dry bites" where no venom is injected. Like many of the country's elapids, brown snake venom is highly neurotoxic.

Other species in the *Pseudonaja* genus are the Speckled Brown Snake (*P. guttata*), Ringed Brown Snake (*P. modesta*), Dugite (*P. affinis*), Peninsula Brown Snake (*P. inframacula*) and Ingram's Brown Snake (*P. ingrami*).

Top: Western Brown Snakes are not as venomous as the Eastern Brown.

Eastern Brown Snake *Pseudonaja textilis*

A mainstay of agricultural areas, the Eastern Brown may be the most regularly encountered dangerous snake along the east coast.

DESCRIPTION: This smooth-scaled snake is variable in colour. From brown, tan or russet (to almost black) as a base, occasionally with a paler head, and perhaps with darker mottling or subtle banding. The underbelly is cream, yellow or light orange, with a pattern of distinct rust-brown or orange splotches. Length rarely exceeds 1.7 m.

BEHAVIOUR: It seeks shelter in sheds and under abandoned debris (sheets of corrugated iron are a favourite). Strictly diurnal, the Eastern Brown assumes a characteristic open-mouthed, S-shaped threat posture before striking. It constricts prey after delivering its bite.

DANGER TO HUMANS: The venom is extremely neurotoxic and the species is well known for its speed and willingness to attack. *P. textilis* is responsible for most Australian snake bite deaths.

AV. SIZE: 1.5 m

HABITAT: Forests, woodlands, grasslands, farms and gardens along east coast (except in extremely arid or wet areas); often lives close to humans

FIRST AID: Pressure immobilisation bandage; antivenom available; seek immediate medical attention

D&D RATING: Very dangerous

Western Brown Snake *Pseudonaja nuchalis*

Also known by its Aboriginal name (Gwardar), the widespread Western Brown Snake shares some parts of its range with the Eastern Brown but is not as venomous.

DESCRIPTION: Like the Eastern Brown is variable in colour. Mostly glossy yellowish-brown or red-brown. May have darker herringbone patterning, or broad dark bands, and a blackish head and neck. Undersurface light yellow or cream with darker orange spots.

BEHAVIOUR: Tolerates more arid habitats than *P. textilis*, including deserts, but also thrives in heavily wooded environments. Like the Eastern Brown, the Western Brown hunts rodents, birds and assorted reptiles (including snakes) in the heat of the day.

DANGER TO HUMANS: The Western Brown is considered the more placid relative of the Eastern Brown, but will readily attack if provoked. While its venom is not as strong, a bite from this snake must be treated with the utmost seriousness.

AV. SIZE: 1.5 m

HABITAT: Forests, woodlands, grasslands, farms and gardens along east coast (except in extremely arid or wet areas); often lives close to humans.

FIRST AID: Pressure immobilisation bandage; antivenom available; seek immediate medical attention

D&D RATING: Very dangerous

Death Adders

Traditionally, only three major death adder species have been recognised. In recent years several more have been described, although their validity is widely disputed. The two species most likely to be encountered are the Common Death Adder (*Acanthopis antarcticus*) and Northern Death Adder (*Acanthopis praelongus*), both described opposite.

Superficially, death adders resemble the classic "viper-like" snake. They are, however, classified as members of the family Elapidae and, despite similar habits, are not "true" adders (Viperidae). Both their behaviour and form demonstrate convergent evolution at work. The death adder is instantly recognisable for its broad, almost triangular, head; short, fat body; huge fangs and thin, rapidly tapering, "grub-like" tail.

Death adders are cryptic snakes that rely on camouflage and ambush tactics to capture prey. They will often partially bury themselves in leaf litter or sand and wait patiently for frogs, reptiles, birds and small mammals to pass within striking distance. To entice nearby animals, the death adder will wriggle its vermiform, or "worm-like", appendage close to its head, inviting the unsuspecting victim to partake of an easy meal. (Such a physiological trait is known as Peckhamian, or aggressive, mimicry.) Once lured, prey is attacked with incredible speed and accuracy.

Despite their fearsome reputation, death adders are reluctant biters and often quite difficult to provoke — these snakes apparently prefer to conserve their venom for prey. Nevertheless, they produce large amounts of highly neurotoxic venom that has killed people in the past. Unlike other elapids, death adders are reluctant to retreat when confronted and therefore pose a problem to any bushwalkers that unwittingly stomp on these well-camouflaged snakes.

Left: A Desert Death Adder.

Common Death Adder *Acanthopis antarcticus*

Common Death Adders are the largest of Australia's death adder species, but are rarely seen due to their idle habits and superb camouflage.

DESCRIPTION: Death adders are short, robust "viper-like" snakes with a broad, flattened (almost triangular) head distinct from the neck, and long fangs. The tail tapers suddenly. Colouring is red-brown to grey with paler bands across the body. The undersurface is pale with dark speckles.

BEHAVIOUR: This is a sedentary ambush hunter that partially buries itself in leaf litter beside pathways used by prey. It wiggles the tip of its tail to lure frogs, reptiles, mice and birds within striking distance.

DANGER TO HUMANS: Because of its superb camouflage, bushwalkers are at risk of standing on this snake and provoking a bite. Its strike is extremely swift, accurate and venomous, and leads to paralysis without proper medical aid.

AV. SIZE: 60 cm

HABITAT: Forests, grasslands and arid areas of New South Wales, Queensland, eastern Northern Territory and southern South Australia

FIRST AID: Pressure immobilisation bandage; antivenom available; seek immediate medical attention

D&D RATING: Very dangerous

Northern Death Adder *Acanthopis praelongus*

This smaller relative of the Common Death Adder is just as lethal, but only occupies areas north of the Tropic of Capricorn.

DESCRIPTION: This short, "fat" snake has a rapidly tapering tail and a broad, flattened (almost triangular) head noticeably distinct from its narrow neck. The body colour is dull orange, red-brown or dark grey with pale grey or yellowish bands, and undersurface pale with dark spots.

BEHAVIOUR: Like the Common Death Adder, this species is a patient hunter that conceals itself in undergrowth and uses its worm-like tail to lure prey. Its fangs have a good degree of rotation compared to other Australian snakes.

DANGER TO HUMANS: Like its cousin, this snake will remain still if approached. A careless foot can easily be bitten. Stout boots, garters and long pants should be worn when hiking through the bush.

AV. SIZE: 50 cm

HABITAT: Range of grassland and forest habitats, north of Tropic of Capricorn

FIRST AID: Pressure immobilisation bandage; antivenom available; seek immediate medical attention

D&D RATING: Very dangerous

Tiger Snakes

Members of the genus *Notechis* derive their common name from the banded, or tiger-like, colouration that may be present on the snake's body. Often, however, this pattern is barely visible and the snake can also be uniformly dark brown or jet black. All tiger snakes are solidly built, with a broad head that is barely distinct from the neck.

Tiger snakes have a somewhat scattered distribution, but generally inhabit the cooler and wetter areas of the continental mainland and are common throughout Tasmania and many of its offshore islands. There are several recognised species of tiger snake, but they are not always easy to identify and are usually distinguished by their geographic range. Some subspecies, like the Chappell Island Tiger Snake (*N. scutatus serventy*), are restricted to one island. Determining the correct species is especially important in the clinical treatment of bites, due to the fact that different species display varying levels of toxicity and therefore require different doses of antivenom.

Generally, tiger snakes live a diurnal existence but have been observed moving across suburban lawns on hot summer nights. They prefer areas close to waterways (for example, river flats, dams and swamps) where they can hunt frogs — the tiger snake's preferred prey. Unfortunately, tiger snake habitat is also appealing to humans as it tends to have agricultural and residential value, so development has caused the snakes' numbers to decline.

Tiger snakes, along with brown snakes, are a frequent cause of serious envenomation in Australia. Their venom has been more extensively researched than any other snake and causes particularly unpleasant effects in humans, including nerve and muscle damage and blood clotting.

Top and opposite: As is the case with so many snake species, colour in tiger snakes is highly variable and is not a reliable method of identification.

Common Tiger Snake *Notechis scutatus*

One of the main culprits of serious snake bite in Australia is the Common Tiger Snake. It often lives close to humans and may enter houses — perhaps resulting in a bite to surprised residents.

DESCRIPTION: The body is smooth and moderately robust and the head is only slightly distinct from the neck. Colour is extremely variable but often olive-grey, red-brown or dark. Bands are commonly present.

BEHAVIOUR: The Common Tiger Snake is active during the day, but will hunt on warm summer nights in search of frogs — its preferred prey. Unlike its Chappell Island relative, this species will bite readily if provoked.

DANGER TO HUMANS: Before antivenom was developed, bite victims had only a 50% survival rate. The Common Tiger Snake has been responsible for a number of deaths in Australia.

AV. SIZE: 1.4 m

HABITAT: Temperate and humid areas of southern Australia, often close to waterways; common in Tasmania

FIRST AID: Pressure immobilisation bandage; antivenom available; seek immediate medical attention

D&D RATING: Very dangerous

Chappell Island Tiger Snake

Notechis scutatus serventy

This subspecies ekes out a harsh existence on cold, windswept Chappell Island, feeding for only a few weeks every year on the chicks of migratory shearwaters. Despite extended starvation, this snake grows larger than its mainland counterpart.

DESCRIPTION: This large, thick-set snake often grows to more than 2 m. Usually glossy black, this subspecies may have pale yellow or brown bands on its body, and a yellow or grey belly.

BEHAVIOUR: About 1500 tiger snakes live on Chappell Island. The snakes hibernate most of the year then gorge themselves on "muttonbird" chicks during the brief nesting season.

DANGER TO HUMANS: Because there is no permanent human population on Chappell Island, these snakes show little fear of humans. Though it is a relatively calm subspecies, any bite should be treated as potentially life-threatening.

AV. SIZE: 1.6 m

HABITAT: Restricted to Chappell Island, Tasmania

FIRST AID: Pressure immobilisation bandage; antivenom available; seek immediate medical attention

D&D RATING: Very dangerous

Black Snakes

Black snakes belong to the genus *Pseudechis*, which includes some of Australia's largest and most familiar venomous snakes. The six known black snake species occupy habitats in all of the country's mainland States and feature prominently in the country's reptile lore. Some species, such as the Red-bellied Black Snake (*P. porphyriacus*), have been branded with an undeserved reputation for ferocity, while others, such as the misleadingly named King Brown Snake (*P. australis*), have rightly earned our respect.

In general, black snakes are not as venomous as other native elapids. Contrary to popular opinion, the bite of the Red-bellied Black Snake is very reluctantly delivered and does not often make humans seriously ill. No documented deaths have been caused by this snake and bites are often "over-treated" with antivenom. Having said this, the species does have the potential to kill — children may be especially susceptible.

The King Brown, or Mulga Snake is the largest, most widespread and perhaps the most dangerous member of Australia's black snake gang. Like the Red-bellied Black Snake, its venom is less potent than that of taipans, tigers, browns and death adders, but it makes up for this by producing the greatest amount of venom of any dangerous Australian snake. The species is also famous for its aggression. When angered, the King Brown Snake will flatten its entire body and strike rapidly. Compounding a successful bite is the snake's habit of latching onto and chewing its victim — thus delivering even more of its ample venom supply.

Other members of the black snake genus are Butler's Mulga Snake (*P. butleri*), the beautifully coloured Collett's Snake (*P. colletti*), Spotted Black Snake (*P. guttatus*) and the Papuan Black Snake (*P. papuanus*). Like all other snakes, they are best not interfered with, in order to avoid a potentially fatal bite.

Top: A King Brown senses "in stereo" — each fork of the tongue collects scent chemicals from the air to indicate the direction of food, a mate or an enemy.

King Brown Snake *Pseudechis australis*

The King Brown Snake is actually a member of the black snake family. Its alternate common name, Mulga Snake, may be preferable as brown snake antivenom is not effective when treating bites from this species.

AV. SIZE: 2 m

HABITAT: Wide variety of habitats, excluding Tasmania and Australia's southern coast

FIRST AID: Pressure immobilisation bandage; antivenom available

D&D RATING: Very dangerous

DESCRIPTION: Australia's second-largest venomous snake is heavily built. It is olive, copper or near-black in colour, and lighter edges on the scales give a reticulated appearance.

BEHAVIOUR: The King Brown Snake hunts a wide variety of prey and is especially partial to other snakes. It is often active at night. To rest it seeks shelter in rabbit burrows and similar places.

DANGER TO HUMANS: This species is responsible for more than 50% of snake bites in the Northern Territory. It hisses loudly and flattens its neck when threatened. It delivers huge amounts of venom (the most of any native snake) and may "chew" victims, thus injecting even more venom.

Red-bellied Black Snake *Pseudechis porphyriacus*

Although this snake has bitten a lot of people, no deaths have yet been recorded. Its taste for frogs has, unfortunately, led to its decline; it is often duped into attacking and eating venomous Cane Toads.

AV. SIZE: 1.8 m

HABITAT: Moist or wet regions of eastern Australia (swamps, lakes, rainforest etc.)

FIRST AID: Pressure immobilisation bandage; antivenom not usually required, but available; seek immediate medical attention

D&D RATING: Dangerous

DESCRIPTION: The glistening, coal black back of this snake contrasts with striking red lower lateral scales (which are not actually the snake's belly).

BEHAVIOUR: The Red-bellied Black Snake is an accomplished swimmer that loves moist, wet areas of eastern Australia where there are plenty of frogs to eat.

DANGER TO HUMANS: Although it has a bad reputation, this snake is a reluctant biter. It flattens its neck and hisses loudly when threatened, but prefers escape to open conflict.

Sea Snakes & Sea Kraits

Some 70 species of sea snake, separated into five groups, are currently recognised worldwide and about half of these species live in Australian waters.

Like terrestrial snakes, sea snakes are cold-blooded, have scales and a forked tongue, and need to breathe air to survive. However, a number of special adaptations have allowed these reptiles to occupy marine environments. A greatly elongated right lung helps them stay underwater for extended periods and controls buoyancy; valves on their nostrils can be held shut while submerged; a paddle-shaped tail provides efficient propulsion; and a gland under the tongue removes excess salt from the body. Sea snakes also slough, or shed their skin, much more frequently than their land-based kin.

Both "true" sea snakes (Hydrophiinae) and sea kraits (Laticaudinae) evolved from terrestrial elapids and share their predatory instincts. All but one species are expert hunters of fish, particularly eels, which they snatch from nooks and crannies of the reef or root out from the bottom sediment. A major point of difference between these two subfamilies is in the way they breed. Kraits are oviparous (egg-laying) and must come ashore to lay their eggs. True sea snakes are viviparous (live-bearing) and give birth to their young underwater.

Sea snake venom is extremely toxic, but they have relatively short fangs (making venom delivery to humans difficult) and deaths from their bites are extremely rare in Australia. Most species are highly curious and show a disarming amount of interest in divers, but generally these reptiles are considered quite peaceful. Every now and then trawler fishers will receive a bite while trying to extract a sea snake from a net.

Top: The streamlined head and closeable nostrils of the Olive Sea Snake.

Banded Sea Krait *Laticauda colubrina*

Around 70 species of sea snake exist world-wide and they are among the most venomous reptiles on Earth. The Banded Sea Krait is a common sea snake of the Indo–Pacific region.

DESCRIPTION: This is a strikingly patterned sea snake, with alternate black and silver-white bands running from its head to the tip of its paddle-shaped tail.

BEHAVIOUR: Small fish, crabs and eels are prey, and are hunted in mangroves and rocky crevices. Comes to the surface to breathe every 60–80 minutes. It lays its eggs on land.

DANGER TO HUMANS: Sea snakes caught in trawl nets present a real risk to humans who try to remove them. These reptiles possess relatively small fangs, but can still inject powerful neurotoxins when they bite.

AV. SIZE: 1 m

HABITAT: Shallow northern waters (less than 10 m deep) around reefs, rocky shorelines and estuaries

FIRST AID: Pressure immobilisation bandage; antivenom available; seek immediate medical attention

D&D RATING: Very dangerous

Olive Sea Snake *Aipysurus laevis*

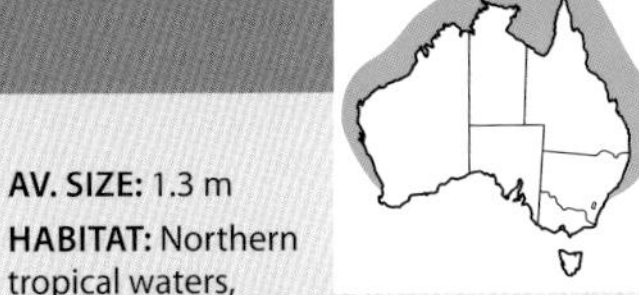

This sea snake often investigates scuba divers on the Great Barrier Reef. It has an extremely curious nature, and wants to discover what is in its realm — perhaps by gently flicking its tongue over a diver. It is not aggressive.

DESCRIPTION: Olive, grey or chocolate brown in colour, with a small head and eyes. Its paddle-shaped tail is often lighter.

BEHAVIOUR: This sea snake hunts fish, crabs and eels around coral reefs in less than 20 m of water and swims in open water. It is a member of one of two families that bear live young underwater. Sexes meet on reefs to breed in winter.

DANGER TO HUMANS: This snake could represent a danger to inexperienced divers who react to its in-your-face curiosity with panic. Remaining calm is essential, and rewarding.

AV. SIZE: 1.3 m

HABITAT: Northern tropical waters, especially around coral reefs

FIRST AID: Pressure immobilisation bandage; antivenom available; seek immediate medical attention

D&D RATING: Very dangerous

Marine Invertebrates

Marine invertebrates are slightly obscure for humans. Lacking any real size or awe-inspiring teeth, they do not immediately spring to mind when we consider what dangers might lurk beneath our waves. Although we may not realise it, Australia's weird and wonderful collection of marine invertebrates includes the most toxic animals on Earth and a number of confirmed killers.

Chief among our dangerous and deadly invertebrates is an underwater army of stinging cnidarians. Though many may seem unlikely candidates as human killers, all cnidarians bear special organs (nematocysts) linked to lethal venom. The most frightening cnidarians are the box jellies, but also dangerous are a number of seemingly innocuous, "plant-like" animals — stinging hydroids, fire corals and pretty sea anemones. Don't let their looks deceive you. These critters are well designed to defend themselves against the unwanted advances of divers and snorkellers.

Also making the list are a handful of molluscs. Thankfully, gastropods such as the cone shell and blue-ringed octopus show little interest in hurting humans and we should always respect such behaviour by leaving them undisturbed; if they went out of their way to cause us harm, surely many more deaths would result from their paralysing toxins.

Echindoerms, such as the Crown-of-Thorns Sea Star (*Acanthaster planci*) and *Diadema* spp. sea urchins advertise their danger with bodies laden with spines. Others, such as the Flower Sea Urchin (*Toxopneustes pileolus*), are more subtle — a flower-like exterior belies this animal's highly toxic capabilities.

Probably the most subtle and seemingly harmless of all are the sponges, which live quiet, filter-feeding lives on the bottom of the sea. Their toxins and shard-like spicules can cause painful injuries to unwary people.

Top to bottom: The wary beauty of a blue-ringed octopus; Sponges come in many shapes and sizes. **Opposite:** The Box Jelly in all its deadly splendour.

Cnidarians

Cnidarians come in a wide variety of shapes and sizes and it may be hard for the average person to believe that many of these creatures are closely related. Included among Australia's "dangerous and deadly" cnidarians are a selection of skin-slicing "true" corals and venomous sea jellies, hydroids, fire corals, box jellies and sea anemones. Concealed within their bodies are some of the deadliest toxins in the world.

Sea jellies are the most familiar dangerous animals in this phylum, and can be divided into two groups. By far the worst of these are the cubozoans, or box jellies — cube-shaped jellies that swim in tropical seas and are discussed in the gravest of terms. Box jellies are further classified in two types — chirodropids (with multiple tentacles trailing from each corner of the "box") and carybdeids (with a single tentacle on each corner). Much less dangerous are the floating scyphozoans, or true jellies, which are the common domed-shaped, "blubber" jellies widely distributed around the continent.

Fire corals and stinging hydroids are colonial hydrozoans — as is the Bluebottle, or Portuguese Man-o-War (*Physalia physalis*). They are made up of thousands of polyps that each perform specific functions in the hydroid's day-to-day life. The polyps that defend the colony can deliver a surprisingly painful sting.

Sea anemones are attractive cnidarians with flower-like, tubular tentacles bearing potent stinging cells. These predatory creatures use these tentacles to capture their prey. In some species, they can cause pain, blistering and respiratory problems in people who touch them.

Top: A White Stinging Hydroid.
Opposite, clockwise from top: Tropical Stinging Hydroid (*Agleophenia cupressina*); Jellyfish; Threespot Humbug (*Dascyllus trimaculatus*) juveniles in an anemone; A Magnificent Hydroid (*Ralpharia magnifica*).

Box Jelly *Chironex fleckeri*

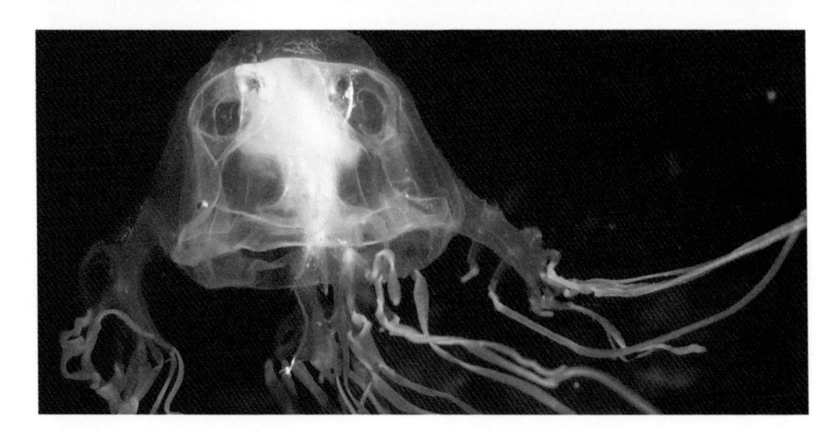

The Box Jelly is one of the world's most venomous organisms and in Australia's tropical latitudes it poses a serious threat to human life between October and April. Contained in the nematocysts of its long, lethal tentacles are toxins that attack the heart, respiratory system, blood and skin cells. Survivors are often left with hideous scars.

DESCRIPTION: The Box Jelly has a semi-transparent, purplish-blue, box-shaped bell with numerous tentacles (up to fifteen) attached in bunches at each of the bell's four corners. The "box" itself may measure up to 20 cm along each side, with tentacles trailing up to 3 m behind adult specimens.

BEHAVIOUR: The Box Jelly represents the peak of sea jelly evolution. It is an active predator capable of propelling itself through the ocean at speeds of 4–5 knots and also has rudimentary eyesight. Death for prey (small crustaceans and fish) is virtually instantaneous and while its venom may appear all out of proportion for its dietary requirements, the Box Jelly has adapted the perfect mechanism for rapid killing in order to avoid any damage to its fragile body.

DANGER TO HUMANS: Since the beginning of the 20th century, approximately 70 people have died after contact with a Box Jelly. Contact with the tentacles causes such excruciating pain that many Box Jelly victims have drowned after lapsing into shock. People have died in less than five minutes following massive envenomation. If you are swimming alone and unprotected in a remote area, and are unlucky enough to blunder into a large Box Jelly, your chances of survival are very slim indeed.

AV. SIZE: 20 cm x 20 cm (bell); 2 m (tentacles)

HABITAT: Tropical coastal waters (beaches, rivers, estuaries)

FIRST AID: Douse tentacles with vinegar; seek immediate medical attention; CPR may be required; antivenom is available

D&D RATING: Very dangerous

Irukandji *Carukia barnesi*

Deriving its name from an Aboriginal language group in the Cairns region, this unseen assassin of tropical seas eluded researchers until the early 1960s when it was first identified as the jelly responsible for the mysterious Irukandji syndrome. Even today very little is known about this tiny jelly's habits or the true properties of its venom.

DESCRIPTION: The Irukandji is a miniscule cubozoan (box jelly) with a transparent body and a single fine tentacle on each corner of its box-shaped bell. Upon close inspection, tiny red dots (the venomous nematocysts) can be seen on the tentacles.

BEHAVIOUR: The habits of *Carukia barnesi* are poorly understood, mainly due to this jelly's small size and near invisibility. Researchers believe the Irukandji adopts a more offshore lifestyle than the Box Jelly, living in deeper reef waters, but it appears in shallower coastal waters during summer months — possibly this movement is caused by prevailing winds and currents. Like the Box Jelly it uses its venom to subdue small fish and crustaceans.

DANGER TO HUMANS: More than 60 people each year are hospitalised with Irukandji syndrome. While the sting itself is not especially memorable, the syndrome that can develop has dramatic systemic effects and is responsible for at least two deaths in Australia. Symptoms become intense 5–30 minutes after contact with the tentacles, and consist of severe and overwhelming pain, muscular aches, vomiting, hypertension and anxiety. These symptoms may last for up to two days. No antivenom has yet been developed, though most victims make a full recovery after a week or so.

AV. SIZE: 1 cm x 1.5 cm (bell); 3–50 cm (tentacles)

HABITAT: Open water and deeper reefs from Childers, Qld to Broome, WA; swept inshore during summer months

FIRST AID: Douse affected area with vinegar; seek immediate medical attention; hospitalisation and powerful analgesics may be required if the syndrome develops

D&D RATING: Very dangerous

Bluebottle *Physalia physalis*

Because so many Australians gravitate towards the country's coastal climes, the Bluebottle, or Portuguese Man-o-War, is often encountered by the country's human population. These familiar floating "jellies" are commonly seen washed up on beaches and countless bathers have suffered stings after bumping into this nasty cnidarian. Fortunately, so far, all have lived to tell the tale.

DESCRIPTION: Not actually a jelly but a colony of different polyps pulling together as a team. The Bluebottle is made up of a blue, gas-filled float with a wrinkled ridge and a well-organised jumble of blue feeding polyps, reproductive zooids and stinging tentacles dangling beneath. The main fishing tentacle, which can stretch for nearly 10 m in some cases, contains the animal's venomous nematocysts.

BEHAVIOUR: The Bluebottle may seem to be an aimless drifter at the whim of wind and tide, but studies have shown that its float may possess some aerodynamic properties. The Bluebottle's movements can be controlled by muscular contractions in the ridge, which tilts to either the left or right. The main tentacle is constantly trawling the water beneath, ready to zap tiny crustaceans and zooplankton.

DANGER TO HUMANS: The Bluebottle has been responsible for human fatalities overseas, but no deaths in Australia. Onshore winds will often cast thousands of Bluebottles upon beaches. A curious Aussie custom is to walk along the beach and pop the gas-filled sac of stranded Bluebottles with the heel of the foot. Since the long stinging tentacle can remain venomous after the Bluebottle's death, this sometimes results in embarrassing, and agonising, entanglements.

AV. SIZE: 2–13 cm (float); 2 m (tentacles)

HABITAT: Coastal waters all around Australia; most common in summer

FIRST AID: Remove tentacles with towel or forceps; immerse affected area in hot — not boiling — water to alleviate pain; see medical attention for pain relief

D&D RATING: Dangerous

Hairy Jelly *Cyanea* spp.

Several species of hairy jelly are widely distributed throughout Australia and the Indo–Pacific region. Colourful names such as Lion's Mane, Hairy Blubber or Sea Nettle are used to describe members of the Cyanea *genus, which are distinctive for their stringy or "hairy" tentacles.*

DESCRIPTION: *Cyanea* species have a flattened, saucer-like, light yellow to brown bell that is divided into eight lobes. Strands of delicate tentacles trail in separate V-like clusters beneath the bell.

BEHAVIOUR: Drifts with currents, feeding on small fishes, crustaceans and plankton.

DANGER TO HUMANS: Contact with *Cyanea* spp. jellies causes an uncomfortable burning sensation, developing into severe pain, which may last for an hour. The nematocysts of dead jellies washed up on beaches can still be triggered and cause injury.

AV. SIZE: 30 cm (bell); 50 cm (tentacles)

HABITAT: Offshore and coastal waters all around Australia

FIRST AID: Remove tentacles and rinse stung area with seawater (do not use freshwater); apply cold pack or wrapped ice; seek medical advice for pain relief

D&D RATING: Treat with caution

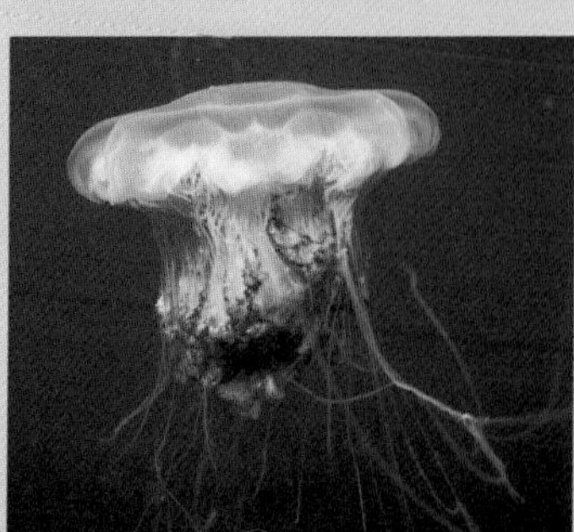

Mauve Stinger *Pelagia* spp.

Found all around the world, Pelagia *species sometimes form huge swarms and this behaviour has been known to put a halt to surf lifesaving competitions in Australia.*

DESCRIPTION: These "stingers" have a symmetrical, semi-transparent mushroom-shaped bell, which displays phosphorescent qualities at night. The bell features "warts" on the upper surface that contain nematocysts. There are sixteen marginal lobes, eight marginal sense organs and eight marginal, hair-like tentacles. Usually pinkish-purple or mauve in colour.

BEHAVIOUR: This open water species drifts with ocean currents and traps small fish, crustaceans and plankton in its tentacles.

DANGER TO HUMANS: Little is known about the venom of this genus, but it causes unpleasant local pain upon contact.

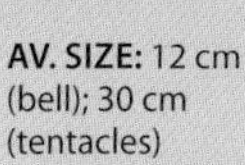

AV. SIZE: 12 cm (bell); 30 cm (tentacles)

HABITAT: Offshore and coastal waters all around Australia

FIRST AID: Remove tentacles and rinse stung area with seawater (do not use freshwater); apply cold pack or wrapped ice; seek medical advice for pain relief

D&D RATING: Treat with caution

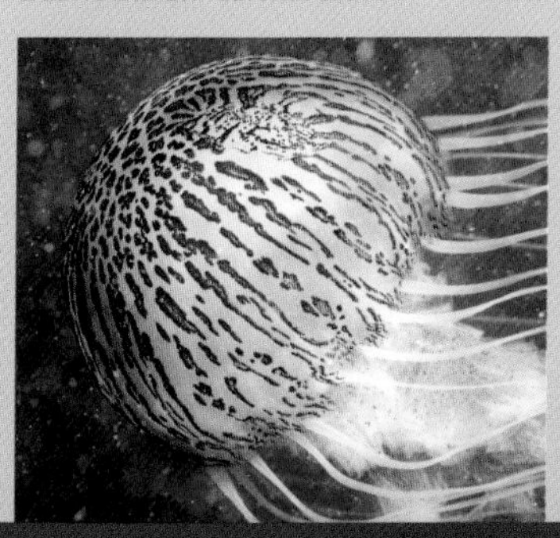

Fire Corals *Millepora* spp.

Despite their appearance and name, fire corals are not "true" corals but venomous animals more closely related to sea jellies.

DESCRIPTION: Fire corals of the genus *Millepora* are often yellow-green or brown in colour with a tough calcareous exoskeleton. They have a superficial resemblance to soft corals or seaweed, and display a variety of growth forms (flower-like, tree-like, feather-like etc.). Fire corals are covered in a layer of surface pores from which fine, hair-like structures (cnidocils) protrude. These trigger stinging cells when touched.

BEHAVIOUR: Fire corals live in colonies in shallow sunlit waters and host symbiotic algae in their tissues. The algae (zooxanthellae) convert sunlight into food, which they share with the fire coral in return for a safe place to live.

DANGER TO HUMANS: Stings from fire corals can cause an itchy rash and pain that may last for several days.

AV. SIZE: Range in size

HABITAT: Tropical and subtropical coral reefs, from central Qld to Shark Bay, WA; warm waters around power station outlets

FIRST AID: Douse affected area with salt (not fresh) water or vinegar; remove residual nematocysts with tweezers, apply ice pack or anaesthetic cream

D&D RATING: Treat with caution

Stinging Hydroids *Lytocarpus* spp., *Aglaophenia* spp. & *Macrorhynchia philippina* spp.

Also known as "fire weed", stinging hydroids look a lot like underwater ferns. They are, however, animals that attach themselves to rocks and possess powerful stings.

DESCRIPTION: Fern-like or feathery in appearance. Various colours including drab green, white or purple. Central stalk from which numerous branching "fronds" radiate. These are covered in rows of tiny polyps.

BEHAVIOUR: Stinging hydroids are often found attached to reef in areas where surge or current brings them a constant supply of plankton to feed on.

DANGER TO HUMANS: Novice divers or snorkellers may be tempted to touch these animals. A painful rash often develops within minutes and may blister.

AV. SIZE: 30–100 cm (high)

HABITAT: Tropical and subtropical shallow reefs, often attached to wharves, pylons and other artificial structures

FIRST AID: Douse affected area with salt (not fresh) water or vinegar; remove residual nematocysts with tweezers; apply ice pack or anaesthetic cream

D&D RATING: Treat with caution

Sea Anemones *Various* spp.

Many people, especially children, find the captivating colours and languid motions of sea anemones almost irresistible. However, many species can sting and there are only a number of creatures equipped to touch these "sea flowers".

DESCRIPTION: Sea anemones come in a range of colours. The anemone has a soft cylindrical body with a foot at the base and an oral disc, with a central mouth, at the top. The oral disc is encircled by a ring of mucous-coated tentacles.

BEHAVIOUR: Anemones are predatory cnidarians that attach themselves to hard substrate or the sea floor via an adhesive foot. The anemone's tentacles are armed with stinging nematocysts that contain toxins deadly to small fish.

DANGER TO HUMANS: *Dofleina armata*, a large sand-dwelling anemone, can cause severe wounds that may take as long as a month to heal.

Above: A large anemone affords vital shelter to one of its family of resident anemonefish, which are protected from the nematocysts.

Above: Some other fish and crustaceans (like this anenome crab) manage to live among the stinging tentacles of a sea anemone.

AV. SIZE: 10 cm

HABITAT: Australia-wide, often in intertidal zones

FIRST AID: Douse affected area with salt (not fresh) water or vinegar; apply ice pack or anaesthetic cream

D&D RATING: Treat with caution

Molluscs

For most people, the word "mollusc" doesn't immediately bring any dangerous or deadly creatures to mind. Quite the opposite. Perhaps you conjure images of oysters drizzled with lemon, barbecued scallops, marinated mussels and grilled calamari. In Australia, we are blessed with an abundance of delectable molluscs that make a significant contribution to the nation's two billion dollar fishing and aquaculture industry. However, several members of the phylum Mollusca (including the ones we eat everyday) have the ability to kill.

Bivalve molluscs, or shellfish (such as oysters and mussels) are filter-feeders that take in nutrients from the surrounding water in order to survive. This lifestyle makes them vulnerable to pollutants, bacteria and toxins, which can accumulate to dangerous levels in their tissues. People who eat these poisonous shellfish can become sick or, in extreme cases, die. A common cause of shellfish contamination is the sudden "blooming" of toxic dinoflagellates — microscopic organisms that attach themselves to algae and are absorbed by bivalves during so-called "red tides". There are at least four kinds of shellfish poisoning, each associated with a different toxin. Perhaps the best known is paralytic shellfish poisoning (PSP), which kills about 300 people around the world each year and has no known antidote.

But it is not just edible shellfish that can cause us harm. Two mollusc genera in particular contain some of the most dangerous toxins in the world. Blue-ringed octopuses (*Hapalochlaena* spp.) and several types of cone shell (*Conus* spp.), described below, have caused fatalities in Australia. Fortunately, these creatures are relatively easy to identify and are only dangerous when handled.

Top: The narrow end of a cone shell forms an opening for extension of the proboscis and the long shell side is where its fleshy "foot" emerges for movement. **Right:** Bivalve molluscs such as oysters and mussels can accumulate toxins to dangerous levels in their tissues.

Cone Shells *Conus* spp.

More than 70 species of cone shell are found around the Australian coast, but only half a dozen are considered dangerous to humans. The animal inside the cone shell is actually a predatory marine snail, harbouring a cache of venomous, barbed harpoons, which it fires from a proboscis that emerges from the narrow end of the cone.

DESCRIPTION: These gastropods live in glossy, intricately patterned, cone-shaped shells. A slit-like aperture runs along the length of the underside, while the blunt end of the cone features a low spire.

BEHAVIOUR: During the day, cone shells bury themselves in sediment, and they emerge at night to feed. Their prey consists of small fish, marine worms and other gastropods (including other cone shells). The cone shell's harpoon is actually a hollow radula tooth dipped in venom. The snail only fires one harpoon at a time, but stores a "quiver" of spare teeth inside a special sac.

DANGER TO HUMANS: Worldwide, cone shells have caused at least fifteen human deaths. The only Australian death occurred in 1935 on Hayman Island — the offending species being the extremely venomous Geography Cone (*Conus geographus*). People with an eye for pretty shells who are not wise to the ways of these deadly gastropods may be tempted to collect cone shells. They should never be handled or placed in pockets as the venomous harpoon can easily penetrate clothing.

AV. SIZE: 10 cm

HABITAT: Australia-wide (but generally tropical) in shallow marine waters, sandflats, mudflats

FIRST AID: Pressure immobilisation bandage; seek immediate medical attention; CPR may be necessary

D&D RATING: Dangerous

Blue-ringed Octopus *Hapalochlaena* spp.

These small, seemingly inoffensive octopuses are one of the world's most toxic creatures, packing enough venom in their golf-ball sized frames to kill a healthy adult human — painfully. Although the blue-ringed octopus is generally thought of as a single species, there are actually several distinct members (at least four) of the* Hapalochlaena *genus.

DESCRIPTION: Blue-ringed octopuses are small — less than a human hand span — and normally display a drab, sandy brown colour, with darker brown or ochre bands marking each of the eight arms and body. Faint blue circles, lines and figures-of-eight are visible on top of these bands. When threatened, however, an octopus undergoes a dramatic colour change. It rapidly turns "electric" as the bands darken and its rings flash a livid blue. This compelling display is caused by pigment cells within the skin (chromatophores), which all cephalopods possess and use to great effect to camouflage themselves in their various habitats.

BEHAVIOUR: Blue-ringed octopuses are common invertebrates found right around the Australian coastline. They are carnivorous hunters that use their venomous saliva to paralyse prey, which mostly consists of crabs and other crustaceans.

Octopus hunting strategy varies according to circumstance. One tactic is to loom over the top of its prey and spray venomous saliva into the surrounding water. Its victim quickly absorbs the toxins, is immobilised and then ripped apart for consumption. Prey is also seized and held tightly by the octopus' muscular arms — aided by rows of suction discs — until a venomous bite can be delivered via its hard, parrot-like beak (located at the junction of its arms). The venom is contained in two salivary glands located above the brain, with a duct connecting these glands to the mouth parts.

DANGER TO HUMANS: The principle component of octopus venom has similar chemical properties to tetrodotoxin — one of the most lethal biotoxins known to science and a poison also found in pufferfishes and some cone shells. The bite itself is little more than a nip and the victim may not immediately experience any pain. However, if enough venom has entered the site, tingling around the face, tongue and lips will be noticed within a matter of minutes. The venom acts swiftly to block the body's nerve impulses

AV. SIZE: 10 cm

HABITAT: Intertidal/subtidal areas, usually around crevices, reefs, rock pools

FIRST AID: Pressure immobilisation bandage; seek immediate medical assistance; CPR may be required

D&D RATING: Dangerous

Top: Flashing bright warning colours, rather than blending in, is a tactic used to alarm. **Above:** Jet-propelled departure.

and more serious symptoms, such as difficulty seeing or breathing, may then follow. Muscle weakness, vomiting and loss of coordination indicate a progression towards total paralysis. If artificial respiration is not performed at this point, the victim may lose consciousness, cease breathing and die.

Mercifully, blue-ringed octopuses are not antagonistic animals. When handled they will flash their warning colours and try with all their might to escape their persecutor's clutches. Usually they will only resort to biting when fully constrained. The genus has caused two deaths in Australia.

Echinoderms

All echinoderms are marine creatures and the phylum is the largest on Earth without any freshwater or terrestrial species. Almost every member of this ecologically important group lives a slow-moving, benthic (bottom-dwelling) existence. The most familiar echinoderms are sea urchins, "starfish" — more accurately called sea stars — and sea cucumbers.

The phylum name Echinodermata is derived from a Greek word that means "spiny skin" and although these creatures are completely non-aggressive towards humans, several species are known for sharp, penetrating spines that can introduce toxins when these animals are handled carelessly.

The sea urchins that cause injury to humans are armed with either long venomous spines, short venomous spines or pedicellariae (small, jaw-like "seizing" organs) that are linked to poison sacs. When handled these spines are easily broken off, sinking deeply into the skin and releasing their venom. The result is burning pain and swelling around the lesion, which often turns blue due to a pigment in the spine. Perhaps the most dangerous sea urchin in the world is the Flower Sea Urchin (*Toxopneustes pileolus*), whose flower-like pedicellariae can cause severe pain and paralysis.

The most infamous echinoderm is a sea star. While the Crown-of-Thorns Sea Star (*Acanthaster planci*) is well known in Australia as a coral killer, it is also protected by long venomous spines that can cause agonising wounds that may take months to heal. It is therefore strongly advised that divers and snorkellers admire the beauty of our coral reefs with their eyes only.

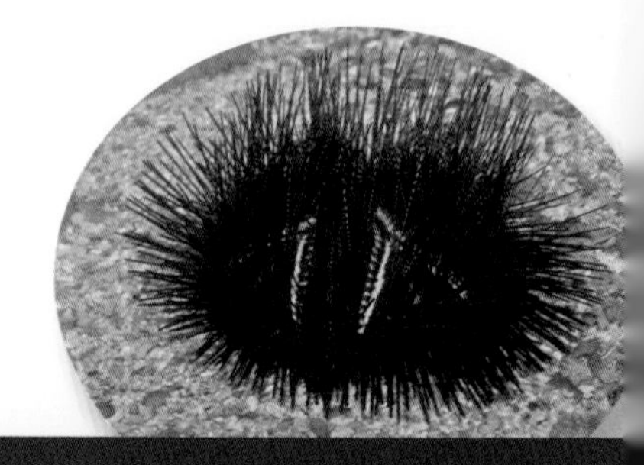

Top: A colourful Flower Sea Urchin (*Toxopneustes pileolus*). **Right:** The sharp spines of a sea urchin.

Crown-of-Thorns Sea Star *Acanthaster planci*

Outbreaks of this tough, invasive species usually signal the beginning of the end for large patches of coral reef, which this species attacks with fabled voracity.

DESCRIPTION: An unusually large purple-blue sea star, with up to 21 arms, covered in long, venomous, crimson-tipped spines.

BEHAVIOUR: A specialist coral feeder, an individual Crown-of-Thorns Sea Star can consume over 5 m^2 of coral every year. It everts its stomach (turns it inside-out) over a 20 cm patch of coral to feed. It is a prolific spawner that is notoriously difficult to eradicate.

DANGER TO HUMANS: Toxins (saponins) in the spines have steroid-like properties and cause agonising pain when they puncture human flesh. Severe discomfort and vomiting may last for hours. Victims may require hospitalisation.

AV. SIZE: 40 cm

HABITAT: Indo-Pacific coral reefs.

FIRST AID: Remove loose spines, but do not attempt to remove deeply embedded spines; immerse affected area in hot — not boiling — water; seek medical attention

D&D RATING: Treat with caution

Sea Urchins *Various* spp.

DISTRIBUTION MAP SHOWS DANGEROUS URCHINS ONLY

These spiny marine grazers come in a variety of shapes and colours. Approximately 80 species are venomous to humans.

DESCRIPTION: Sea urchins consist of a round, globular shell (a "test") that radiates spines, which may be either short and stubby, long and slender, or covered in pedicellariae.

BEHAVIOUR: Muscles control spines on the underside of the urchin's test, moving it slowly along the seabed. A mouth (Aristotle's lantern) at the bottom of the test allows urchins to scrape off and eat algae, seaweed and other encrusting animals.

DANGER TO HUMANS: *Diadema* species are known for their needle-like spines that effortlessly penetrate skin; collecting these urchins for their roe may result in injury.

AV. SIZE: 10-20 cm

HABITAT: Temperate and tropical seas

FIRST AID: Remove loose spines, but do not attempt to remove deeply imbedded spines; immerse effected area in hot — not boiling — water; seek medical attention

D&D RATING: Treat with caution

Right: Fire Urchin (*Asthenosoma varium*).

Sharks

The human ideal of the "shark" thrives at the very core of our subconscious mind and provokes a generalised and deep-rooted unease when most of us enter the ocean. It spans cultures and demographic divides and is an ideal that resists all attempts at rationalisation. It is an oft-quoted statistic that we have a greater chance of being struck by lightning than being attacked by a shark (30 times greater in fact). As it stands the humble mosquito is, quite literally, a million times more deadly, yet it cannot wrest away even an ounce of the morbid romanticism we attach to the shark.

Approximately 170 shark species inhabit Australian waters. Of these, only three — the Bull, Tiger and White Shark — have been regularly implicated in human fatalities and therefore must be considered dangerous. Entering the water near a shark does in no way guarantee an attack. Even the most dangerous of sharks, the Great White, has been safely observed by free-swimming divers.

It is a fact that most people survive shark attacks and that people killed during an attack are only occasionally eaten. But attacks on humans still occur, at a rate of roughly one per year. To reduce the odds of a shark attack even further, avoid swimming at first light or after sunset — these are the times when sharks are likely to be close to shore in feeding mode. Steer clear of river mouths, deep gutters, sudden drop-offs and murky water. Don't swim in canals or other areas that Bull Sharks are known to inhabit. If you are spearfishing, try not to remain in the water too long after spearing a fish.

Top: A Scalloped Hammerhead Shark (*Sphyrna lewini*). **Opposite:** A diver swims safely with Oceanic Whitetip Sharks (*Carcharhinus longimanus*).

The White Shark — commonly known as the Great White or, less commonly these days, the White Pointer — is the largest and most feared predator in Australian waters. Its infamy as a dangerous animal (in the tabloid media at least) is unrivalled, and if looks could kill it would certainly live up to its terrible reputation. Almost in defence of its own character, the White Shark continues to defy sensationalism. What little is known of this fish's true behaviour hints at a species altogether more complex than the simple "killing machine" we demand that it be.

Top: A cruising White Shark. **Opposite, top to bottom:** Pectoral fins are shaped like plane wings, for lift and planing; The upper jaw protrudes on attack.

DESCRIPTION: The White Shark is a huge fish with a cone-shaped snout, black eyes and near symmetrical profile. The tail is lunate, or crescent-shaped, with similarly sized upper and lower caudal fins. The dorsal and pectoral fins are large. The White Shark's streamlined, torpedo-like body displays clearly marked countershading, with a grey dorsal area and snow-white underbelly. Both the upper and lower teeth are broad, triangular and serrated. Like many other sharks, the White Shark's dentition is in a constant state of flux — lost and broken teeth are constantly replaced by rows of new teeth in "conveyer-belt" fashion.

BEHAVIOUR: The White Shark is an uncommon, warm-blooded pelagic fish that roams widely throughout the

AV. SIZE: 4–4.5 m

HABITAT: Coastal and offshore temperate and sub-tropical waters; patrols islands, deep gutters and open ocean habitats

FIRST AID: Control visible bleeding by applying direct pressure; treat for shock; seek immediate medical assistance

D&D RATING: Very dangerous

world's temperate and subtropical seas. In Australia they have been recorded from southern Queensland, round the south of the continent to north-west Western Australia, with concentrated populations in the Great Australian Bight. Juveniles feed on fish, squids, rays and other sharks, but as individuals reach maturity they become specialist hunters of marine mammals, especially pinnipeds.

The White Shark's attack strategy can be likened to mugging. Essential to its success is the element of surprise — a requirement for capturing such fast-moving and agile prey as fur-seals. The shark's countershading also plays an important role in attacks. When viewed from above, its dark dorsal colours blend in with the surrounding waters, while from below the white belly creates minimum silhouette against the sunlight. Typically, the shark will launch a single massive strike from below and then retreat to a safe distance until its victim bleeds to death. The White Shark can generate a tremendous amount of speed and propulsion with its muscular tail, often resulting in spectacular airborne attacks.

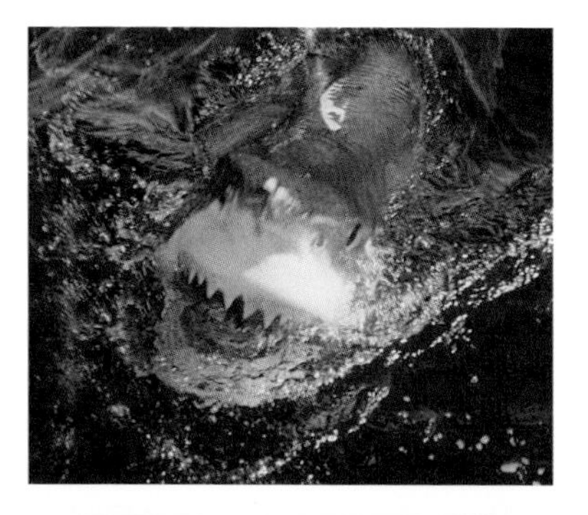

DANGER TO HUMANS: To date, the White Shark has claimed the lives of 40 people in Australian waters, injured 27 more and attacked, but not harmed, a further 26 people. For such a large and capable predator these numbers are somewhat surprising. Unlike seals, humans provide the White Shark with an easy — in most cases stationary — target in the water, and many experts speculate that attacks on people are often either a case of "mistaken identity" or "taste-testing". There is no denying that from below the silhouette of a surfer on his/her board bears an uncanny resemblance to a seal, which satisfies the former theory. The latter can be explained by a shark's lack of sensory appendages (like our hands or feet). Unfortunately for us, it would seem that the only way for a shark to ultimately determine a foreign object's edibility is to take a bite.

Requiem Sharks

Requiem, or whaler sharks, belong to the family Carcharhinidae, which includes such dangerous and deadly representatives as the Tiger Shark (*Galeocerdo cuvier*), Bull Shark (*Carcharhinus leucas*) and Bronze Whaler (*Carcharhinus brachyurus*). These three species are the best known members of the family and all are large fish capable of killing humans, but they perhaps deserve vindication for simply being so well-equipped to "do what sharks do".

Many requiem shark species have regular contact with humans. On the Great Barrier Reef, Whitetip Reef Sharks (*Triaenodon obesus*) and Blacktip Reef Sharks (*Carcharhinus melanopterus*) are relatively harmless sharks commonly seen by divers. Grey Reef Sharks (*C. amblyrhynchos*) are the star attraction of many shark dives, but are known to become aggressive during feeding frenzies — such feelings, however, are well advertised by their exaggerated threat displays.

Around more southern surf beaches, the Bronze Whaler (*C. brachyurus*) is another requiem shark that regularly moves among swimmers — though many people may not be aware of their presence. They are known to hunt fish in the surf zone, and vigorous splashing, spearing fish or wearing brightly coloured jewellery has been known to catch their attention. In these situations, most attacks are hit-and-run affairs and usually not fatal. The shark, after snatching at a hand or foot, usually flees the scene swiftly. This may explain the high number of non-fatal attacks ascribed to "whalers in general" (60 in total).

In deep offshore waters, the Blue Shark (*Prionace glauca*) and Oceanic Whitetip (*Carcharhinus longimanus*) roam widely in their search for prey. You have to be tough to survive the open ocean, and both these large species are known for directing aggressive behaviour towards humans.

Top: A Blue Shark.
Far left: Effective counter-shading of a Whitetip Reef Shark. **Left:** Blacktip Reef Shark on a reef.

Bull Shark *Carcharhinus leucas*

Arguably the most dangerous shark in the world, due to its imposing physique, close association with humans, ability to survive in freshwater habitats and extremely testy nature — the Bull Shark is probably responsible for more human fatalities than it is officially credited with in the International Shark Attack File.

DESCRIPTION: The Bull Shark is a stout, heavyset whaler with a short, blunt snout and small eyes. Its body is uniformly grey above and pale underneath. The second dorsal fin is about one-third the height of the first. No interdorsal ridge is present. The upper teeth are triangular and serrated (for shearing), while the lower teeth are more finely pointed (for gripping).

BEHAVIOUR: A solitary hunter, the Bull Shark adapts to a wide range of environments in warm waters around the world. This species spends a lot of time close to shore, is tolerant of murky conditions, and is known to haunt canals (such as those of Queensland's Gold Coast) and the upper reaches of rivers where other sharks cannot survive. The Bull Shark is a highly territorial species and said to have the highest testosterone levels of any animal in the world.

DANGER TO HUMANS: The Bull Shark has killed ten people in Australian waters, but the true figure is likely to be higher. Being a large, aggressive shark that stakes out territories in densely populated areas, it poses a threat to bathers in many parts of the country. Because the Bull Shark is a specialist of murky water, there may be no sign of an impending attack. Like all sharks, this species hunts mostly at dawn and dusk.

AV. SIZE: 2.5 m

HABITAT: Various coastal and estuarine habitats, including freshwater and brackish reaches of rivers; also thrives in canals, harbours and lakes

FIRST AID: Control visible bleeding by applying direct pressure; treat for shock; seek immediate medical assistance

D&D RATING: Very dangerous

Tiger Shark *Galeocerdo cuvier*

The mighty Tiger Shark, one of the world's largest predatory fishes, is a confirmed "maneater" that ranks second only to the White Shark in terms of the danger it poses to humans. The Tiger Shark's impressive bulk, coupled with its innate curiosity and scavenging nature, has resulted in 23 fatal attacks in Australia's tropical and warm temperate seas. The species is famous for its indiscriminate feeding — research autopsies have revealed a range of inanimate objects in its stomach (plastic bottles, tin cans and shreds of rubber). Such behaviour has earned the Tiger Shark the unflattering title of "garbage can of the sea".

DESCRIPTION: The Tiger Shark is so named for the dark spots and vertical bars that pattern its flanks. These are especially prominent in juveniles and gradually fade as the shark matures. The body is greenish-grey in colour and pale beneath. The species has large eyes, a blunt, wedge-shaped head that allows for sudden maneuvering, and a long upper caudal fin. An interdorsal ridge is present. The Tiger Shark's teeth, contained in a large maw, display a characteristic "cockscomb" design and can easily shear through the hard shells of marine turtles.

BEHAVIOUR: The Tiger Shark is primarily a nomad. It is wide-ranging throughout the world's oceans, undertaking seasonal migrations from the tropics to more temperate waters in summer (and returning in winter). Tiger Sharks are mostly nocturnal hunters, lurking on the deeper fringes of reefs during the day and exploring shallower waters at night. Prey includes fish, sharks, squids, sea birds, rays, marine turtles, carrion and various odds and ends of questionable nutritional value. Tiger Sharks have been known to gather in large numbers at known breeding sites of sea birds and marine turtles. Raine Island, off the Cape York coast, attracts large numbers of Tiger Sharks each October and November when Green Turtles arrive there to breed.

DANGER TO HUMANS: The Tiger Shark maintains a consistent vigil in warm oceans around the world — it is a particular menace in the Hawaiian Islands — and should be treated with the utmost respect. Many attacks occur in murky water, where the shark's ability to detect low-frequency vibrations may attract it to human commotion. Tiger Sharks are aggressive and confident fish, with the habit of slowly circling and nudging potential prey prior to an attack. Unlike many shark species, the

AV. SIZE: 3.5 m

HABITAT: Tropical and warm temperate waters; frequents reefs, bays, river mouths, deep channels and harbours

FIRST AID: Control visible bleeding by applying direct pressure; treat for shock; seek immediate medical assistance

D&D RATING: Very dangerous

Tiger Shark will consume its human prey. Indicative of their presence around Queensland's beaches, Tiger Sharks are the main shark species captured on baited drum lines, with more than 6000 sharks caught over the past two decades.

Opposite and above: The deep body and striped marking of the Tiger Shark.

Bronze Whaler *Carcharhinus brachyurus*

Although the lay person would have a lot of trouble accurately identifying one of these sharks in the wild, the Bronze Whaler has nonetheless cemented its place in Australia's beach folklore. The "bronzie" is often cited as the reason behind many lifeguard alarms. The term itself is used by many non-scientists and seems to cover most species of whaler shark.

DESCRIPTION: The Bronze Whaler receives its common name from the coppery sheen that is usually present on its body, which is a brown or greyish colour and white beneath (exhibiting countershading that is common to all whaler sharks). The snout is moderately slender and pointed. The upper teeth display narrow, "hooked" cusps, while the lower teeth are more oblique and finely serrated. These whalers have unusually large pectoral (side) fins with dusky to black tips and dusky tips on their pelvic fins. No interdorsal ridge is present.

BEHAVIOUR: Often seen, with other whaler species, feeding on bait schools near the surf zone, Bronze Whalers also patrol deep waters around offshore islands. They feed on pelagic and bottom-dwelling fish, as well as squids, stingrays and other sharks.

DANGER TO HUMANS: The Australian Shark Attack File attributes three human fatalities to this shark. Because the Bronze Whaler is a large whaler that is known to regularly patrol surf zones, it should be considered dangerous.

AV. SIZE: 2 m

HABITAT: Cooler waters from Moreton Bay, Qld, south to Jurien Bay, WA

FIRST AID: Control visible bleeding by applying direct pressure; treat for shock; seek immediate medical assistance

D&D RATING: Dangerous

Oceanic Whitetip *Carcharhinus longimanus*

The Oceanic Whitetip is a large solitary shark that roams the world's deep oceans far offshore and is rarely encountered by humans. However, it is often quick to arrive on the scene of any mid-ocean disasters (plane crashes or sinking vessels) and was a prime suspect for human fatalities during WWI and WWII.

DESCRIPTION: The Oceanic Whitetip is so named for the distinctive, mottled white tips of its first dorsal, pectoral, pelvic and caudal fins. The dorsal fin is widely rounded, while the long, wide-swept pectoral fins have an obvious paddle shape. The body is stocky, grey to dusky in colour with a light underside. Occasionally a darker "saddle" may be visible between the first and second dorsal fins. The nose is short and rounded. Upper teeth are broad, triangular and serrated. Bottom teeth are narrower and only serrated towards the tip.

BEHAVIOUR: This species is known for its brazen and opportunistic nature (a necessity of its open water habitat where any foreign object could represent food and therefore demands investigation). Remoras and Pilotfish are usually seen in tow and the shark often follows pods of Pilot Whales, scrounging leftover meals of squid. This strong "following instinct" often has it tailing tuna boats, seeking fish and dumped refuse.

DANGER TO HUMANS: Its bold, bullying behaviour and sheer size command respect. Although divers have safely observed the Oceanic Whitetip in the wild, it is an unpredictable shark, prone to "feeding frenzies", and has killed humans in the past.

AV. SIZE: 3 m

HABITAT: Offshore in tropical and warm temperate seas

FIRST AID: Control visible bleeding by applying direct pressure; treat for shock; seek immediate medical assistance

D&D RATING: Dangerous

Marine Fish

About 4400 described fish species live in Australian waters, with the greatest levels of diversity occurring on tropical reefs. The country's deadly and dangerous fish fauna can be somewhat crudely categorised by the type of danger they present to people. On the one hand, there are fish that can inflict severe musculo–skeletal trauma — sharks and moray eels for example. On the other hand are the toxic fishes — the scorpion fishes, puffers and a number of other species. These fish carry venom in their spines or toxins in their flesh which, when introduced to the human bloodstream, can cause significant ailments. Some of these venomous fish, the stingrays for example, have the ability to cause serious soft tissue wounds in addition to their toxins.

For most people, the marine fish of greatest concern around Australia are the sharks. The vast majority (about 95%) of Australia's shark species are harmless but because of the media hype surrounding shark attack they seem to loom larger in our psyche than many other marine fish. In the previous chapter we discussed the so called "big three" — the White, Tiger and Bull Sharks. We saw how they formed a deadly triumvirate that has maintained its position at the top of the International Shark Attack File list.

Less sensational, though no less deadly, are the toxic fish. Prominent among this group are the Scorpaeniformes (lionfish, stonefish and their ilk) and the stingrays — close relatives of the shark. These fish lack the aggression of large, toothy predators and instead use venom, delivered through stout spines, as a form of defence. Other members of this spike brigade are the various catfishes, surgeonfish, flatheads and rabbitfish that are not especially lethal but can inflict deep, venomous puncture wounds if handled or cornered. Finally, there are the fish whose danger is essentially gastronomic — the Tetraodontiformes (pufferfishes) and ciguatera-carrying species whose flesh should never be eaten, but all too often is consumed with disastrous consequences.

Top: A toxic pufferfish. **Above:** A stingray.
Opposite: The strikingly patterned Common Lionfish.

Surgeonfish *Acanthurus* spp.

All surgeonfish bear a sharp scalpel-like blade on each side of their tail just in front of the caudal fin. They receive their common name from the precise and highly efficient lacerations these blades inflict.

DESCRIPTION: Mostly surgeonfish are moderately small fish with "pouting" mouths and long dorsal and anal fins that run almost the entire length of the body. Many species display lurid warning colours and intricate patterns all over the body.

BEHAVIOUR: Surgeonfish are non-aggressive herbivores that feed on detritus and algae around reefs.

DANGER TO HUMANS: The surgeonfish's blades normally lie flat but will spring to life when the fish becomes alarmed. Overly stressed fish, usually those caught in nets, will thrash their tails and cause deep cuts that remain painful for several days.

AV. SIZE: 15–30 cm

HABITAT: Tropical and subtropical reefs

FIRST AID: Remove victim from water; apply pressure to stop bleeding

D&D RATING: Treat with caution

Left: The scalpel-like blade on the side of the tail. **Below:** Displaying warning colours.

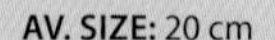

Rabbitfish *Siganus* spp.

Rabbitfish also go by the derisively ironic name of "happy moments". These prickly little characters sport a considerable armoury of venomous spines — thirteen dorsal spines, four pelvic spines and seven anal spines.

DESCRIPTION: Smooth-skinned, moderately small, bream-like fish. *Siganus lineatus* is a common species with numerous gold dashes and dots over its head and body and a larger yellow patch near the tail.

BEHAVIOUR: Rabbitfish are herbivores and are usually found in schools. Their common name comes from their flighty response when confronted by divers.

DANGER TO HUMANS: The venomous spines of these fish cause immediate and long-lasting pain. They must be handled with a great deal of care and attention if they become an accidental capture while fishing.

AV. SIZE: 20 cm

HABITAT: Tropical and subtropical and temperate seas, around reefs, estuaries and rivers

FIRST AID: Immerse affected area in hot — not boiling — water; seek medical advice

D&D RATING: Treat with caution

Right: A Masked Rabbitfish (*Siganus puellus*).

Bullrout *Notesthes robusta*

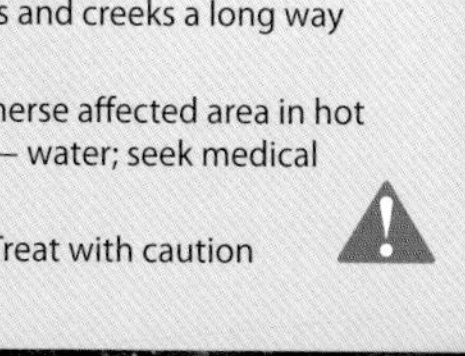

Also known as the Kroki for the peculiar grunting noise it emits, the Bullrout leads a fairly indolent existence on the bottom of estuaries.

DESCRIPTION: The Bullrout has a large, spiny head, cod-like appearance and fifteen stout, venomous dorsal spines. It has a blotchy or marbled appearance and varies in colour from pale to dark brown.

BEHAVIOUR: The Bullrout is known to venture long distances upstream where it can maintain its sluggish lifestyle in slow-moving or still waters.

DANGER TO HUMANS: Very little biochemical assessments have been made of the Bullrout's venom, but it is known to cause instantaneous and sickening pain. Like the stonefish, its strong dorsal spines can penetrate rubber-soled footwear. Injuries and lymph glands can remain tender for some time.

AV. SIZE: 15 cm

HABITAT: Estuaries and lowland rivers; also dams, weirs and creeks a long way upstream

FIRST AID: Immerse affected area in hot — not boiling — water; seek medical advice

D&D RATING: Treat with caution

Gurnard Perch *Neosebastes* spp.

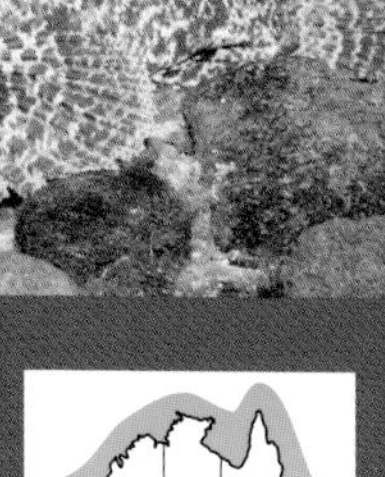

Several gurnard perch species are found around Australia and are considered excellent table fish with firm, white flesh. All have highly venomous dorsal spines and should be carefully handled when captured.

DESCRIPTION: The gurnard perch has a relatively large, gargoyle-like head with a huge mouth and slender body. Like other scorpionfish the gurnard perches have wide pectoral fins and long dorsal spines.

BEHAVIOUR: This bottom-dwelling genus is known for its greedy behaviour. Fish rest on the bottom, often at depths of up to 500 m, and gulp down any suitably sized prey that swims past.

DANGER TO HUMANS: This genus has caused two deaths in Australia. In one case, a Flinders Island man was stung by a dead fish stored in his refrigerator. Its venom was thought to have accelerated a prior medical condition.

AV. SIZE: 30 cm

HABITAT: Rocky reefs and sandy substrates to a depth of 500 m

FIRST AID: Immerse affected area in hot — not boiling — water; seek medical advice

D&D RATING: Dangerous

Estuarine Stonefish *Synanceia horrida*

This cumbersome, gargoyle-like "rock with fins" definitely won't win any underwater beauty pageants, but it deserves admiration for its superbly adapted physiology and frightening defensive arsenal. It is the most venomous fish in the world — each of its stout dorsal spines bears two venom glands and delivers an unbelievable amount of pain to anyone unfortunate to stand on it.

DESCRIPTION: The word "ugly" is almost always used to describe the Estuarine Stonefish. It is a bulky and ungainly animal with a large underslung mouth and a body covered in warty, greenish-brown, mottled skin that perfectly imitates a rock. The dorsal fins are noteworthy for their huge size in comparison to the body. The tail is relatively small and tapered.

BEHAVIOUR: The Estuarine Stonefish makes its living as an ambush predator in warm, shallow waters. Using its strong dorsal fins to partially bury itself in sand or mud, it lies dormant on the bottom for long periods and waits patiently for its meals to swim past. When a small fish or crustacean ventures too close, the fish explodes into action — engulfing prey in its huge mouth. The stonefish's speed is disarmingly quick and it will also chase prey over short distances. Unlike most fish, *Synanceia horrida* can survive many hours out of water.

DANGER TO HUMANS: The Estuarine Stonefish's excellent camouflage makes it a danger to anyone walking in shallow water without protective footwear. Even so, its hard, needle-sharp spines have been known to penetrate rubber-soled shoes. Stonefish venom has powerful neurotoxic properties.

AV. SIZE: 35 cm

HABITAT: Shallow waters from Moreton Bay, Qld to Houtman Abrolhos Group, WA

FIRST AID: Immerse affected area in hot — not boiling — water; administer analgesic; seek immediate medical aid; antivenom available

D&D RATING: Very dangerous

Common Lionfish *Pterois volitans*

This hardy, fearless and completely mesmerising fish is hugely popular among saltwater aquarists. Novice fish-keepers, however, should not be deceived by the Common Lionfish's Siren-like allure. Most of their fins conceal venomous spines that can puncture careless hands and induce serious illness.

DESCRIPTION: The Common Lionfish is a perch like member of the Scorpaenidae family — a group of marine fish well-known for their venomous spines. Strikingly patterned with zebra-like maroon or brown and white bars across its body, this species is also instantly recognisable for its long dorsal fins and showy, pennant-like pectoral fins. The adipose, caudal and anal fins are often spotted. Fleshy, "leaf-like" tentacles sprout from around the mouth and eyes.

BEHAVIOUR: Common Lionfish are commonly seen swimming in a leisurely fashion around reefs and rocky bommies in water up to 50 m deep. They are often encountered in pairs. The fleshy tentacles, which resemble algal growths, may serve as a kind of disruptive camouflage to break up the outline of the fish's head and thus help conceal it from predators. Common Lionfish hunt at night. Like other scorpionfishes, they appear lethargic but can strike rapidly — gulping down small fishes and crustaceans. The species is also prone to cannibalism and has been observed "herding" prey into rocky corners with its pectoral fins.

DANGER TO HUMANS: This species seems to show little fear of humans, often approaching divers with their dorsal spines tilted forward in defence. Thirteen dorsal spines, three anal spines and two pelvic spines are attached to venom glands.

AV. SIZE: 30 cm

HABITAT: Reefs and rocky substrates from south-west WA around the tropical north to southern NSW

FIRST AID: Immerse affected area in hot — not boiling — water; administer analgesic; seek immediate medical aid

D&D RATING: Dangerous

Stingrays Various species

Stingrays are the largest venomous fish in Australian seas and are represented by at least 50 different species. Each is equipped with a long, serrated barb — located at the tail end of the vertebral column — which is used for defence. Death by stingray is an extremely rare event in Australia, though injuries are more common than many people would suspect. More often than not it is the barb itself, and not the venom, which causes the most damage.

DESCRIPTION: The stingray is the flat relative of the shark and is immediately familiar for its rounded body shape and oversized pectoral fins (often thought of as "wings"). The stingray's eyes are positioned on top of the head, and display a "sunken" appearance due to adjacent breathing spiracles. The mouth is located on the underside of the body.

BEHAVIOUR: Stingrays are docile bottom-dwellers that bury themselves in sand and mostly remain motionless. However, they can quickly speed away with a shimmering of their strong pectoral fins if threatened. Stingrays feed on a variety of molluscs, shellfish, worms and crustaceans.

DANGER TO HUMANS: Australia's three recorded deaths have all been inflicted by a barb directly entering the heart. The most famous of these was the death of celebrated "wildlife warrior" Steve Irwin in 2006. But the majority of incidents occur when rays are stood on in shallow water — lacerations to the ankle are the most common injury.

AV. SIZE: 30–120 cm (body disc)

HABITAT: Australia-wide, mostly in shallow water less than 30 m deep

FIRST AID: Immerse affected area in hot — not boiling — water; administer analgesic; seek immediate medical aid

D&D RATING: Treat with caution

Moray Eels *Gymnothorax* spp.

Because morays are occasionally involved in attacks on humans, they are often given a bad rap in popular literature — and their appearance probably doesn't help. In most circumstances, however, it is our own behaviour and lack of understanding about these creatures' habits that prompts morays to bite.

DESCRIPTION: Morays of the genus *Gymnothorax* are moderately large eels with beady eyes, razor-sharp teeth and powerful jaws. Most exhibit spots or ornate markings on their mucous-covered body, which is muscular and laterally compressed. Morays have a single tube projecting from each nare (nostril), a small circular gill on each side of the head and a long, unbroken dorsal fin that runs from the back of the head to the tail and "links" with an anal fin that runs all the way up the underbelly.

BEHAVIOUR: Moray eels spend their days hiding in cracks and crevices of the reef, emerging at night to hunt fish (including other eels), molluscs and crustaceans. They are often seen poking their heads out of dark alcoves with their mouths agape. This looks threatening but is a way of helping the eel take in enough water to breathe.

DANGER TO HUMANS: Attacks commonly occur around dive sites where "tame" morays are regularly fed as an attraction. This practice quickly accustoms them to humans and, like many animals, in such situations they can become pushy and lacerate stray hands and fingers.

AV. SIZE: 1–1.5 m

HABITAT: Tropical, temperate and subtropical reefs and weed-covered bottoms to 150 m

FIRST AID: Control bleeding with direct pressure to wound; treat for shock

D&D RATING: Treat with caution

Pufferfish Various species

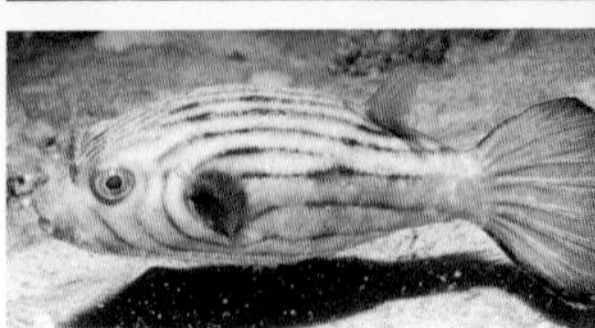

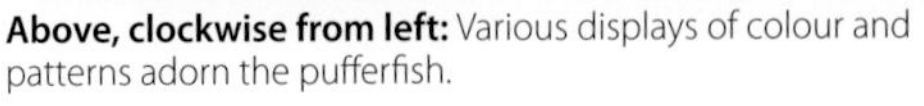

Above, clockwise from left: Various displays of colour and patterns adorn the pufferfish.

Pufferfish, also known as toadfish or "toadies", belong to the Tetraodontidae family. This family contains many species with exceptionally toxic internal organs and skin, which are nonetheless considered a delicacy in some Asian countries — most notably Japan. In fact, it is the very presence of this poison that makes the puffer so appealing as a dish.

DESCRIPTION: Most puffers are slow-swimming fish, often sporting colourful patterns, with tiny gill slits and fused parrot-like teeth. They are usually smooth, though the related porcupinefish are covered in long, prickly scales. Most puffers lack a first dorsal fin.

BEHAVIOUR: When threatened, all puffers can inflate their bodies to balloon-like proportions by gulping water into the stomach. In addition to their nerve toxins and unwelcoming threat displays, some puffers have a highly cantankerous nature. The Ferocious Puffer (*Feroxodon multistriatus*) has been known to chomp at the fingers and toes of unsuspecting swimmers. In some cases, digits have been completely bitten off.

DANGER TO HUMANS: In 2006, two sailors working on a Chinese ship caught and consumed a pufferfish while they were moored off Dampier, Western Australia. One man died of respiratory paralysis after being transferred to Karratha hospital, while the other survived. A third crew member fell ill after attempting artificial respiration and accidentally ingesting residual nerve toxin. Such a gruesome scenario demonstrates the inherent risks of eating these lethal fish.

AV. SIZE: 10–90 cm

HABITAT: Shallow tropical and temperate seas; some species enter brackish and freshwater habitats

FIRST AID: If victim is awake, induce vomiting; seek immediate medical aid

D&D RATING: Very dangerous

Ciguatera

Ciguatera poisoning has been a common affliction of the Indo–Pacific region for many centuries. As many as 50,000 cases of this unpleasant food-borne sickness are reported worldwide each year. Though incidents can occur almost anywhere in Australia, it is a particular health concern in the tropics.

DESCRIPTION: Ciguatoxin, the toxin that causes ciguatera poisoning, originates in microscopic organisms called dinoflagellates. *Gambierdiscus toxicus* is the primary ciguatoxic agent and is passed up the food chain to predatory fish that are commonly prepared for the table.

BEHAVIOUR: *Gambierdiscus toxicus* dinoflagellates attach themselves to algae that are eaten by small fish species, which are in turn eaten by larger species. By this stage, dangerous levels of the toxin may have accumulated. Although most tropical predatory fish can carry ciguatoxin, certain species are regularly implicated in cases of human poisoning. These include the Chinaman, Red Bass, Paddletail and Spanish Mackerel. Although not customarily eaten in Australia, moray eels are also known to harbour dangerous levels of ciguatoxin.

DANGER TO HUMANS: Ciguatoxin is an extremely hardy and virtually undetectable poison. It is tasteless and resistant to freezing, heating or processing of fish. The toxin causes gastrointestinal and neurological effects in humans but does not harm its carriers. Symptoms appear 1–24 hours after eating poisonous fish and include numbness or tingling around the mouth, nausea, diarrhoea, joint and muscle pain, extreme itchiness, a reversal of hot/cold sensations and hallucinations.

Below, clockwise from top left: Ciguatera is microscopic and can be passed on from fish-to-fish; Coral Trout (*Plectropomus leopardus*); Chinamanfish (*Symphorus nematophorus*); Red Emperor (*Lutjanus sebae*).

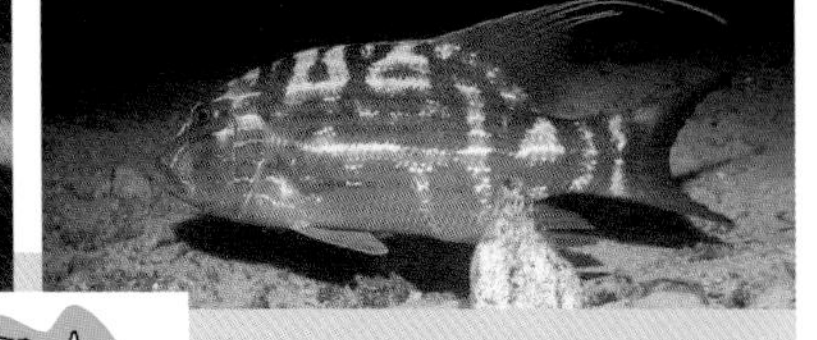

AV. SIZE: Mostly in fish larger than 50 cm

HABITAT: Mostly tropical, but can occur anywhere in Australia

FIRST AID: Induce vomiting if patient is awake and only within 1–3 hours after eating; keep well hydrated; seek medical aid

D&D RATING: Dangerous

Spiders

Australia is home to approximately 70 spider families and more than 2000 described species. However, the gaps in our spider knowledge are considerable and some experts estimate that the total number of species may be as high as 10,000. There is also a fair amount of general misinformation and hysteria attached to Australian spiders. Only two species — the Red-back (*Latrodectus hasselti*) and the Sydney Funnelweb (*Atrax robustus*) — have caused human deaths. An antivenom is available for both species and no deaths have been recorded since these came into use. Not a single person has died from the bite of a native spider since 1979. We do however still need to be aware of a number of species and the potential effect of their bites.

Spiders are broadly classified into two infraorders — Mygalomorphae, or primitive spiders, and Araneomorphae, or modern spiders. Mygalomorphs are stout, ground-dwelling spiders whose ancestors toughed out life in humid Permian forests more than 250 million years ago. Araneomorphs are web-weaving spiders that evolved in the Jurassic Period alongside increasing numbers of flying insects. Apart from their lifestyles, one major difference between these families is the way their fangs operate. Mygalomorphs have large fangs that strike downwards like pick-axes. By contrast, araneomorphs have smaller fangs that come together in a pincer-like fashion.

As a general rule, Australia's mygalomorphs are the spiders you need to keep an eye on. Known for their aggression, this family includes the funnelwebs (about 40 species), mouse spiders (eight species), trapdoor spiders (81 species) and Australian tarantulas (seven species). Araneomorphs are more timid, though many members will defend themselves when escape appears unlikely. The Red-back Spider is the most dangerous member of this family, which also includes orb weavers, wolf spiders, huntsmen spiders and the controversial "flesh-eating" White-tailed Spider (*Lampona cylindrata*).

Top: A Funnelweb Spider (*Kiama lachrymoides*).
Opposite: Bold colour of a Red-back Spider.

Sydney Funnelweb Spider *Atrax robustus*

With B-movie looks and an aggressive attitude, the male Sydney Funnelweb Spider is arguably the planet's most dangerous arachnid and a chief protagonist in Australia's dangerous and deadly folklore. It is also one of the country's largest spiders, satisfying the "big, black and deadly" imagery that lurks in the dark corners of every arachnophobe's mind. The male is six times more venomous than the female — unusual in the spider world — and is more often encountered. However, no deaths have been recorded in Australia since an antivenom became available in 1980.

DESCRIPTION: The male Sydney Funnelweb is a relatively large, plum-coloured or black spider with a glossy carapace, spiny limbs and decidedly "creepy" appearance. Compared with the female, it has a relatively small abdomen, covered in velvet hairs, with four long spinnerets (silk-spinning organs) protruding from the end. Four book lungs are also visible on the underside of the abdomen. The eyes are arranged in a tight cluster and are set above a pair of massive paraxial (downward-pointing) fangs, which slot like pocket knives into serrated fang grooves. Another prominent feature on the male is a tibial spur, located on each second leg, which is used to restrain the female's fangs during mating.

BEHAVIOUR: Female Sydney Funnelweb Spiders are long-lived, often surviving for more than ten years, and spend most of their time in humid, silk-lined subterranean burrows. Most males mature in 2–4 years and spend their entire adulthood (6–9 months) above ground searching for mates. A spider either digs its own burrow or occupies rocky crevices, natural cavities in root systems or spaces among the foundations of houses. A burrow normally has two entry–exit points built into the main tube, creating a Y-shaped or T-shaped lair. A mess of silk trip lines decorate these entrances and any contact with these strands will very quickly see the spider emerge

AV. SIZE: 4 cm (body length)

HABITAT: Gullies of wet sclerophyll forests and rainforests; moist soil in suburban gardens; Newcastle to Nowra, west to Lithgow

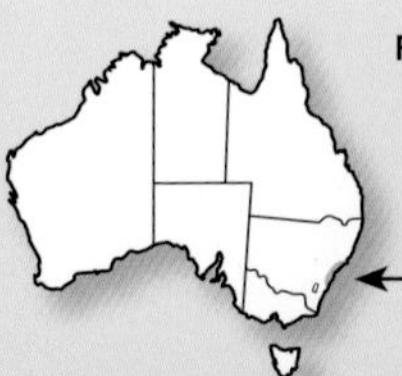

FIRST AID: Pressure immobilisation bandage; seek immediate medical assistance; antivenom available

D&D RATING: Very dangerous

from hiding to challenge its trespasser. Beetles, cockroaches, snails, and millipedes make up the bulk of its diet, though small vertebrates (frogs, skinks and even mice) are also eaten.

DANGER TO HUMANS: The Sydney Funnelweb Spider has killed thirteen people in Australia, but no lives have been lost since the introduction of antivenom in 1980. The spider displays a characteristic threat posture when disturbed and is extremely pugnacious — rearing up with its forelegs held high overhead and fangs dripping with atraxotoxin. Interestingly, primates are the mammals most susceptible to this nerve toxin. Encounters with male funnelwebs are most likely during summer and autumn when they wander suburban gardens in their quest for love. Gardeners are particularly at risk, as digging in the soil can occasionally unearth these animals.

Opposite and above: The threat display of the male funnelweb shows off the spider's impressive fangs.

Red-back Spider *Latrodectus hasselti*

Despite its small size and inoffensive nature, the female Red-back Spider continues to feature prominently in any discussion of Australia's dangerous and deadly animals. While the female is extremely venomous — the male's fangs are too tiny to penetrate human skin — no fatalities have been recorded in Australia since an antivenom became available in 1956. Nevertheless, the Red-back Spider continues to haunt the dark recesses of the Australian psyche. In fact, most people entering an outback dunny will inevitably bring this spider to mind. Such superstition, however, is also reasonably grounded. It would be of little comfort to know that, in all likelihood, a Red-back Spider will be hiding in such a cosy habitat. Fortunately, it is extremely unlikely to bite unless disturbed.

DESCRIPTION: The female Red-back Spider is about the size of a small grape and is generally black, occasionally dark brown, with a smooth, glossy carapace and an overall delicate or "spindly" appearance. Although completely black forms are found, this spider's most distinguishing feature is, of course, the red stripe that runs along the top of its abdomen. Rarely, this stripe will be an orange or pink colour. Commonly, another red "hourglass" marking appears on the underside of the abdomen.

BEHAVIOUR: Red-back Spiders favour dry, sheltered habitat, which is provided by any number of artificial structures in close proximity to humans. The Red-back web is a haphazard entanglement of fine, but extremely sturdy silk that consists of a funnel-shaped upper retreat and central snare where sticky trap threads are connected vertically to ground attachments. When an insect blunders into these trap threads, the ground attachments snap and the insect is propelled up towards the snare. Humans not only provide habitat for this spider, but also attract its prey — cockroaches, ants, mosquitoes and moths. If a female is not ready to mate, she will also capture and consume any male Red-backs that come courting. Red-backs also steal prey items from other spiders' webs.

DANGER TO HUMANS: Like other araneomorphs, the Red-back Spider is not known for its aggression. When threatened, its preference is to curl up into a ball, drop to the ground

AV. SIZE: 1 cm (female); 3 mm (male)

HABITAT: Australia-wide, except alpine areas and extremely arid deserts; common in built environments (letterboxes, outdoor toilets, sheds etc.)

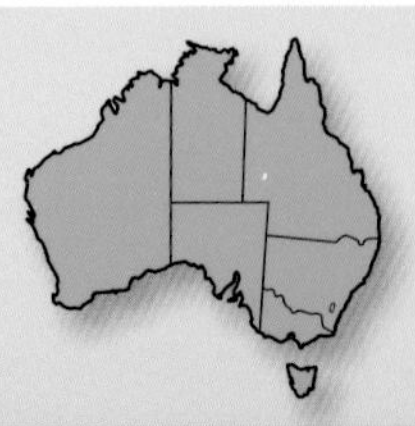

FIRST AID: Ice packs for pain; seek medical assistance; antivenom available

D&D RATING: Dangerous

Note: Do NOT use compression bandage

and scuttle for safety. Although about 2000 bites are reported each year, the vast majority occur when a spider is accidently handled when cleaning up around the home or when it is inadvertently trapped against a human hand or body and has no other means of escape. Annually, about 250 cases require antivenom.

The Red-back and her overseas sisters — the Black Widow of the United States and the Katipo of New Zealand — possess a potentially life-threatening venom known as alpha-latrotoxin, which attacks the human nervous system and is characterised by extreme pain. However, because of the Red-back's small fangs, the initial bite may go unnoticed. Reactions vary from person to person, but a burning sensation may be felt a few minutes later, followed by muscle weakness, "migratory" pain, nausea, tremors and localised sweating. Pressure immobilisation bandages should never be applied as the serious effects of venom take hours, or even days, to develop and constrictive bandages will only increase the victim's pain.

Opposite and above: With a prominent red stripe on her abdomen, the female Red-back Spider is one of Australia's most distinctive arachnids.

Mouse Spiders *Missulena* spp.

Eight species of mouse spider live in Australia. Male mouse spiders are highly venomous trapdoor spiders that are known for their pugnacious personalities. Females are reclusive, excavating elaborate, silk-lined subterranean burrows with two entrances, up to 30 cm deep.

DESCRIPTION: All mouse spiders are squat, stocky arachnids with huge fangs and bulbous chelicerae (the structures bearing the fangs). Females and males of the same species show strong dimorphism (difference between the sexes in body shape and colouring). Males often display the strong warning colours of the natural world — bright red, yellow and blue. Females are generally dark and uniformly coloured and markedly larger than males. Unlike other mygalomorphs, whose eyes are arranged in tight "clusters", the mouse spider's eyes are spread out.

BEHAVIOUR: Like other mygalomorphs, female mouse spiders live underground. Being trapdoor spiders, some species make a lid or door over the entrance to the burrow. Compared to males, female mouse spiders are sluggish and docile creatures. Males, by contrast, lead a more nomadic existence, digging shallow burrows when required. Unlike other mygalomorphs, males roam during the day and may be seen moving about on open ground, usually in winter and especially after rain.

DANGER TO HUMANS: Mouse spiders are aggressive and will bite hard when provoked. Despite their toxicity, only one severe case of envenomation has been recorded in Australia. Typical effects of a mouse spider bite include local pain, profuse sweating and vomiting. Though these effects usually do not last more than a few hours, all mouse spider bites should be treated seriously.

AV. SIZE: 1–3 cm (body length)

HABITAT: Various habitats, including suburban gardens, and often close to waterways

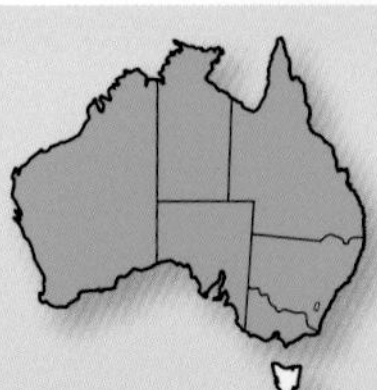

FIRST AID: Pressure immobilisation bandage; seek medical assistance; antivenom available

D&D RATING: Dangerous

Australian Tarantulas *Selenocosima* spp.

Also known as "whistling" or "barking" spiders, Australia's native tarantulas recently attracted attention as a boutique pet in many Australian cities. Ironically, these large spiders try to avoid humans and are arguably the country's most aggressive arachnids. Australian tarantulas are more closely related to Old World (African) tarantulas. New World tarantulas (the "classic" tarantulas of the Americas) are considered more docile.

DESCRIPTION: Australian tarantulas are heavy-bodied, "tarantula-like" spiders that are among the largest Australian spiders. They are represented by two different families and come in a range of colours from light fawn to grey-brown. Their fine covering of hair also imparts a velvet sheen. A prominent feature of these spiders' anatomy is a set of impressive fangs, which can measure up to 10 mm long.

BEHAVIOUR: Australian tarantulas dwell in burrows in the ground, though they can climb any vertical surface with ease. Because of their size and powerful venom, Australian tarantulas take a wide variety of prey, including frogs, lizards, various crawling insects and other spiders. When disturbed they will "stridulate", rubbing their spiny palps against a special organ located at the base of the fangs, and produce a whistling sound — hence their common name.

DANGER TO HUMANS: Keeping one of these spiders as a pet is a bad idea for two reasons — they are naturally reclusive (you will rarely see the spider) and they hate being disturbed. Thankfully, only a handful of bites have been reported and no deaths recorded.

AV. SIZE: 5 cm (body length)

HABITAT: Widely distributed from sandy deserts to rainforests in Queensland, New South Wales, Western Australia and South Australia

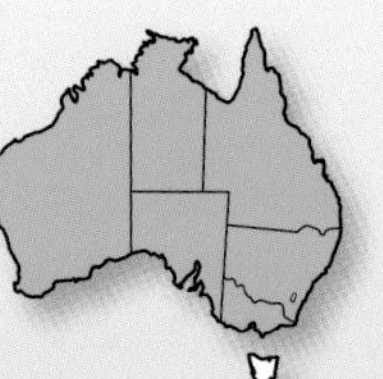

FIRST AID: Wash site; seek medical attention if reactions are severe

D&D RATING: Treat with caution

Insects

Insects are the undisputed overlords of the invertebrate world. For the most part, they go about their daily lives undetected and successfully occupy almost every available habitat on planet Earth. The total number of described insect species numbers more than one million, but very few of these are harmful to humans. Many in fact are a great benefit to us — providing an unnoticed, but vital, cog in the machinery of life. They pollinate our plants and crops, break down plant and animal matter and provide an endless supply of protein to other animals.

Australia's most obvious dangerous and deadly insects can be split into two groups — ants, bees, wasps (order Hymenoptera) and mosquitoes (order Diptera). Some ants, bees and wasps bear a venomous ovipositor at the end of the abdomen, which unleashes a painful sting in all humans and a much more serious anaphylactic response in allergic individuals. Bull ants, paper wasps and European Honeybees (*Apis mellifera*) have all caused deaths in Australia.

Mosquitoes are the most dangerous insects in Australia — and the world for that matter — and transmit a number of crippling diseases. Perhaps the most serious of these is dengue fever, outbreaks of which are presently localised in Queensland's tropics.

Not so dangerous but nonetheless irritating, are several species of hairy caterpillar, including the "processionary" larvae of the Bag-shelter Moth (*Ochrogaster lunifer*). Their poisonous hairs and venomous spines cause uncomfortable, itchy rashes and allergic reactions.

Top: Common Paper Wasps (*Polistes humilis*) at nest. **Opposite:** A bull ant showing its long, straight, serrated mandibles.

Bees, Ants & Wasps

Communal ants, wasps and bees are the perfect role models for social societies. Leading thoroughly organised, altruistic regimes, these widespread and beautifully adapted insects display an innate understanding of the words "team" and "work" and demonstrate a healthy respect for nature — giving as much back to the environment around them as they take out.

Ants, wasps and bees are all closely related. Together these insects comprise the suborder Apocrita, which is part of the large order Hymenoptera. It includes some of the most advanced insects on Earth and a multitude of species that prove both useful and troublesome to humans in everyday life. Nearly every Australian would be familiar with the introduced European Honeybee (*Apis mellifera*), having witnessed its work in parks and gardens around the country and enjoyed the fruit of its labour on their toast. Although it is an introduced species, this bee is single-handedly responsible for about 80% of our crop pollination and many farms would suffer without its presence. It is an equally crucial pollinator of our native flowering plants.

Many people would also be familiar with the stings of these insects, which are delivered by an ovipositor — an egg-laying organ that in some species has been modified into a venom-bearing "needle". Encounters with native paper wasps (*Polistes* and *Ropalidia* spp.), whose nests are often built on the outsides of houses, are not easily forgotten. Unlike the European Honeybee, which can sting only once, wasps and stinging ants can deliver multiple stings and pose a serious threat to people with allergies to their venom. Generally, Australia's most aggressive wasps are social wasps. Solitary wasps, the mud-daubers and spider wasps, are venomous but often appear too busy to bother humans.

Top: Native paper wasps at their nest.
Right: Bull ants are among the most primitive of all ants.

Bull Ant *Myrmecia* spp.

Ants of the genus Myrmecia *are among the most primitive of all ants and more than 99% of all species are endemic to Australia.*

DESCRIPTION: All *Myrmecia* species are extremely large ants with long, straight, serrated mandibles and big eyes. Many species are reddish with darker abdomens.

BEHAVIOUR: Fast-moving foragers, worker bull ants have keen vision and are well known for their extremely aggressive posturing and nasty sting (located on the tip of the abdomen).

DANGER TO HUMANS: Bull ants are common insects and sting many Australians each year. The sting is comparable to that of a wasp and has been known to kill people with hypersensitivity to venom.

AV. SIZE: 13–36 mm

HABITAT: Southern areas of Australia

FIRST AID: Ice packs to reduce swelling; seek immediate medical attention for allergic individuals

D&D RATING: Treat with caution

Imported Red Fire Ant *Solenopsis invicta*

First discovered in Brisbane in 2001, this small, introduced ant poses a serious social, economic and environmental threat. The name "fire ant" comes from the intense burning sensation that is felt after an attack.

DESCRIPTION: Head and body are a copper-brown, and the abdomen is darker. This is a small ant with different sized individuals in a colony, that can easily be confused with small native ants. It builds dirt nests with no obvious entry or exit holes. Established nests are mound-shaped.

BEHAVIOUR: The Imported Red Fire Ant is known for its ferocity, particularly near the nest. Swarming attacks can be an extremely distressing experience.

DANGER TO HUMANS: Fire ants are aggressive and share habitat with humans who have not yet become culturally attuned to identifying these pests. Stings cause an itchy, burning sensation and after a few hours blisters form at the site of each sting.

AV. SIZE: 2–6 mm

HABITAT: Restricted to South-East Queensland, between Brisbane and Ipswich; often nests in open areas (such as lawns, roadsides and pastures) *

FIRST AID: Ice packs to reduce swelling; seek immediate medical attention for allergic individuals; do not puncture blisters

D&D RATING: Treat with caution

* **REPORT FIRE ANT SITES TO THE DEPARTMENT OF PRIMARY INDUSTRIES**
Email: callweb@dpi.qld/gov.au

European Honeybee *Apis mellifera*

Unlike wasps, the European Honeybee's sting is barbed. It can therefore sting only once, but the attached gland will continue to pump venom minutes after the bee has detached itself (and flies off to die).

DESCRIPTION: This is the familiar introduced bee, with a hairy, golden brown body with black stripes across its abdomen and dark legs. It has large dark eyes and transparent wings.

BEHAVIOUR: It constructs colonial nests of wax (secreted from its abdomen) in a honeycomb pattern. Foraging bees that are returning to the nest display ("dance") in order to indicate locations of nearby flowers. These bees are important pollinators of native plants.

DANGER TO HUMANS: European Honeybees will defend their nests against threats but foraging bees are not aggressive. Bees collecting pollen close to the ground (e.g. on clover) are often stood on by humans and will sting.

AV. SIZE: 13–17 mm

HABITAT: Australia-wide

FIRST AID: Remove sting immediately with a sideways scraping action, never attempt to "pull" stings out; seek immediate medical help where anaphylaxis is likely

D&D RATING: Treat with caution

European Wasp *Vespula germanicus*

First discovered in Tasmania in 1959, the European Wasp is the scourge of alfresco dining in southern parts of Australia and an unwelcome environmental pest.

DESCRIPTION: Prominent yellow and black colouration, with long, transparent wings, black antennae and mostly yellow legs. Black markings on the abdomen are arrow-shaped.

BEHAVIOUR: The European Wasp is a social species that constructs concealed nests in walls, ceilings or in the ground. It will eat anything from meat to soft drinks, and fearlessly invades barbecues and picnics.

DANGER TO HUMANS: The European Wasp is a particularly aggressive species and will form swarming attacks near the nest.

AV. SIZE: 12–15 mm

HABITAT: Tasmania, Victoria, ACT, south-eastern New South Wales and wetter regions of South Australia

FIRST AID: Ice packs to reduce swelling; seek immediate medical attention for allergic individuals

D&D RATING: Treat with caution

Paper Wasps *Polistes* spp.

These communal, or "social", wasps are aggressive defenders of their nests and will attack as a group, chasing and repeatedly stinging people who accidently, or deliberately, come into contact with them.

DESCRIPTION: Paper wasps come in a range of colours. The Common Paper Wasp (*Polistes humilis*) is a tan-coloured, slender wasp with long, thin wings and some yellow markings on the face.

BEHAVIOUR: These wasps build "papery", cone-shaped nests — comprised of numerous individual hexagonal cells — that hang from a stalk under foliage or the eaves of buildings. Only female wasps posses a sting.

DANGER TO HUMANS: Most attacks occur when people accidently brush against nests while walking in the bush, or when attempting to remove or destroy nests around the home. Stings are immediately painful and may cause an allergic reaction in hypersensitive people.

AV. SIZE: 10–15 mm

HABITAT: Australia-wide, including urban areas

FIRST AID: Ice packs to reduce swelling; seek immediate medical attention for allergic individuals

D&D RATING: Treat with caution

Spider Wasps *Cryptocheilus* spp.

These wasps are often mistakenly referred to as "hornets" (no true hornet species lives in Australia). Spider wasps are solitary hunters that are often seen dragging huge huntsmen spiders along the ground.

DESCRIPTION: The most common Australian spider wasp is *Cryptocheilus bicolor* — a large wasp with orange wings and legs and a wide orange band encircling its abdomen.

BEHAVIOUR: Spider wasps are common in suburban gardens and are often seen on the ground digging in soft sand and soil or hauling paralysed spiders back to their burrows. They move quickly with "jerky" motions.

DANGER TO HUMANS: Solitary wasps are generally not as aggressive as communal wasps. Nevertheless, their sting can be just as painful, and hypersensitive people are at risk of developing an allergic reaction.

AV. SIZE: 10–35 mm

HABITAT: Australia-wide; common in suburban gardens during summer

FIRST AID: Ice packs to reduce swelling; seek immediate medical attention for allergic individuals

D&D RATING: Treat with caution

Mosquitoes

The mosquito is the most dangerous animal in the world by a wide margin. This tiny insect has remained a consistent killer of humans for many thousands of years and the toll it takes on human life is profound. The World Health Organisation estimates that a child dies of malaria every 30 seconds. It is one of the most insidious diseases known to science and a major health concern with destructive socio-economic impacts. Endemic malaria was eradicated from Australia in 1981, but the country remains a home to several mosquito species that transmit some unpleasant diseases of their own.

Australia accommodates more than 300 mosquito species. Distinguishing these species is almost impossible without the use of a magnifying lens, but all mosquitoes share a common life cycle and anatomy. Like all insects, mosquitoes have a body made up of a head, thorax and abdomen. All are small (usually less than 10 mm), have six legs, two wings and a proboscis that conceals piercing and sucking mouthparts. Females generally lay batches of eggs in stagnant fresh water. These eggs hatch into aquatic larvae ("wrigglers"), which, over the following 1–2 weeks, undergo four separate moults before reaching a floating pupa ("tumbler") stage. Two days later the pupae metamorphose into adult mosquitoes that live for only a couple of weeks.

Both male and female mosquitoes feed on nectar, but females will suck blood in order to obtain the necessary protein for egg development. By sucking blood, female mosquitoes become vectors for a number of hideous viruses and parasites that survive in the blood of people and animals. In Australia, the most important mosquito-borne diseases are dengue fever, Ross River fever, Australian encephalitis and Barmah Forest virus.

Top: It may be small, but the mosquito is the most dangerous animal in the world.

Dengue Fever Mosquito *Aedes aegypti*

Dengue fever is Australia's most dangerous mosquito-borne virus. Queensland is currently the only Australian State that experiences dengue outbreaks (which have proved fatal).

DESCRIPTION: The females of *Aedes aegypti*, which carry dengue fever, have inconspicuous white markings, banded legs and a characteristic sitting position (low and parallel to surfaces with hindlegs raised).

BEHAVIOUR: This introduced, domestic mosquito lives and breeds around houses in tropical Queensland. Generally, bites occur indoors during daylight hours.

DANGER TO HUMANS: It breeds in artificial containers (pot-plant dishes, discarded tyres, bird baths etc.). *Aedes aegypti* depends on humans to complete its life cycle. Install fly screens around your home, empty any garden containers of still water and wear insect-repellent while outside.

AV. SIZE: 5 mm (unfed)

HABITAT: Lives in houses and shaded gardens in tropical Queensland

FIRST AID: See a doctor for blood tests as soon as symptoms (fever, flushed skin, intense headache, joint and muscle pains) appear; stay well hydrated and rest

D&D RATING: Dangerous

Ross River Fever Mosquito *Aedes & Culex* spp.

Named after Ross River in Townsville, where the first case was diagnosed, several mosquito species, living all over the country, are capable of transmitting this virus.

DESCRIPTION: The *Aedes* and *Culex* genera are difficult to identify without the use of a magnifying lens. Unlike *Aedes* spp., *Culex* spp. have a high sitting position with all legs resting on the surface.

BEHAVIOUR: *Aedes* spp. are generally responsible for transmitting Ross River Fever in tropical and coastal areas. *Culex* spp. are more often found in inland Australia and are nocturnal blood-suckers.

DANGER TO HUMANS: Although it is not fatal, Ross River Fever can cause debilitating, fever-like symptoms that may, in rare cases, last up to a year.

AV. SIZE: 4–10 mm (unfed)

HABITAT: All Australian States

FIRST AID: See a doctor as soon as symptoms (fever, flushed skin, intense headache, joint and muscle pains) appear; stay well hydrated and rest

D&D RATING: Dangerous

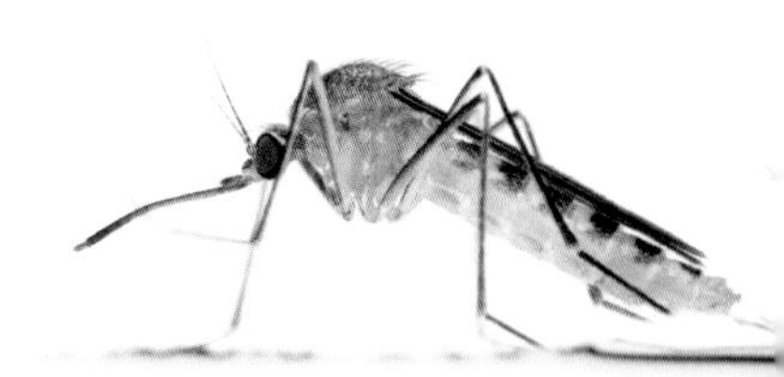

Other Arthropods

All arthropods are distinguished by their segmented bodies, jointed appendages and hard external skeletons (exoskeletons). This is the largest phylum in the animal world and includes all insects, arachnids and crustaceans. However, for the purposes of this book, "other arthropods" refers to a handful of dangerous and deadly animals that are not described in sections dealing specifically with insects or spiders.

Probably the most dangerous of these arthropods are the parasitic ticks, fleas, lice and mites, which feed on the blood of humans and other animals and help pass a range of disease-carrying organisms onto their hosts. The most well known of these arthropods is the Paralysis Tick (*Ioxodes holocyclus*). Contained within its saliva is a nerve toxin that has claimed the lives of twenty people in Australia. While an antivenom has been developed, ticks continue to kill thousands of domestic animals every year.

Less dangerous arthropods are the country's various centipedes and scorpions. Centipedes are commonly encountered in the bush and people's backyards and do pack a powerful punch. Tropical species are often more potent, but their venom, while awfully painful, is not life-threatening. Scorpion stings are poorly documented in Australia and even though two fatalities have been blamed on these iconic arachnids, the reliability of the evidence surrounding these cases is disputed. Nevertheless, scorpions do dish out the odd sting. Smaller members of the family Buthidae should definitely be avoided.

Top: Giant Centipede (*Ethmostigmus rubripes*) attacking a small snake. **Right:** Scorpions should be avoided.

Centipede *Ethmostigmus & Cormocephalus spp.*

Although no one has ever died from centipede venom in Australia, their bite is strong, immediately painful and may last for up to 48 hours.

DESCRIPTION: Reddish-brown, elongated arthropods with flattened, segmented bodies and multiple pairs of legs — always an odd number from 15–191. There are sharp, curved forcipules (poison claws) under the head and modified legs at the tail end for grasping prey.

BEHAVIOUR: Centipedes are common nocturnal hunters that live under rocks, in fallen logs and within soil and leaf litter.

DANGER TO HUMANS: People often come into contact with centipedes while gardening or camping. Be wary when collecting dead wood for fires and always check shoes, socks, gardening gloves etc. that have been left out overnight.

AV. SIZE: 10 cm (with larger specimens found in the tropics)

HABITAT: All habitats, Australia-wide

FIRST AID: Seek medical attention for pain remediation and possible treatment for secondary infection

D&D RATING: Treat with caution

Paralysis Tick *Ixodes holocyclus*

The female Paralysis Tick is highly underrated as a dangerous and deadly creature. People generally associate ticks with dogs, but they have been responsible for at least twenty human fatalities in Australia.

DESCRIPTION: This is a tough, relatively hard-bodied arthropod. Legs and mouthparts are orange-brown, and the body is cream to dark grey, with "sunken" lines running along the back. Engorged ticks may swell to the size of a small grape.

BEHAVIOUR: Ticks wait patiently on foliage for host animals (either wild or domestic) to brush past. Ticks will then attach themselves and begin to feed, introducing a strong neurotoxin after 3–5 days.

DANGER TO HUMANS: Most people have a mild allergic reaction to tick bite, but in some cases (usually children) this may lead to symptoms of paralysis.

AV. SIZE: 3–4 mm

HABITAT: Wet and dry forests and grasslands along coast east of the Great Dividing Range

FIRST AID: Remove parasite promptly; use curved scissors/forceps to lever tick out, being sure to remove mouthparts; antivenom available; seek medical attention for allergic individuals

D&D RATING: Dangerous

Mammals

While many kinds of flesh-eating mammal do survive in Australia, there are none that even come close to the heft of a lion, tiger or bear. In fact, there is only one native carnivore with enough size to trouble a human being — and it does from time to time — the Dingo.

Excluding *Homo sapiens*, the most dangerous mammal in Australia is actually man's best friend, *Canis lupus familiaris* — the domestic dog. Nearly three million dogs are registered in Australia and, compared with any other animal in this book, they bite a lot of people — 30,000 patients present with bite injuries annually, but the total number of bites is likely to be much higher. Its close relative, the Dingo, is much less dangerous (in a statistical sense at least) but has generated a lot of press for itself. Although it has managed to escape the hysteria heaped upon it during the 1980s, its legend as a dangerous animal lives on through the occasional attack. For the most part, this is the fault of humans. On Fraser Island, resident Dingoes are often fed scraps and encouraged by doting tourists. Such behaviour is irresponsible — it only conditions these wild animals to associate humans with food and makes them more antagonistic.

Next in line behind the dog and Dingo are the more unlikely suspects — introduced mammals such as the Asian Water Buffalo (*Bubalus bubalis*) and the Feral Pig (*Sus scrofa*). Asian Water Buffalos are formidable beasts and the bulls in particular should not be trifled with — the species has killed at least one person in Australia. Feral Pigs also present a danger, mostly to people who hunt them for sport. Armed with sharp tusks, and weighing more than 100 kg, an enraged razorback is a worthy adversary for any human or pig dog.

Finally, there are our native bats. In 1996, a new strain of lyssavirus — a disease similar to rabies — was found in several flying-fox and microbat species. The disease is transmitted to humans through the bite of these bats and has caused two known deaths to date. However, since less than 1% of bats carry the virus, and many people do not come into contact with these flying mammals, the risk of contracting the disease is extremely low.

Opposite: A rarely seen Black Dingo from the wilds of central Australia.

Dingo *Canis lupus dingo*

PURE DINGO
HYBRID DINGO

Australia's largest terrestrial carnivore has attracted considerable notoriety since its introduction to the country 3000 years ago.

DESCRIPTION: Lean, medium-sized dog with pointed ears and "wolf-like" features. Its most familiar colour form is ginger/golden. May also be black, white or tan. Underside lighter. Interbreeding with domestic dogs has led to a variety of colour forms.

BEHAVIOUR: Naturally shy Dingoes live in territorial packs and hunt alone or in groups. They are intelligent, opportunistic predators that eat a wide variety of foods.

DANGER TO HUMANS: Fraser Island Dingoes share habitat with tourists who often feed these wild animals. They may act aggressively once food is expected and may view humans (particularly children) as prey or territorial competitors. In 2001, a nine-year-old boy was attacked and killed by two Dingoes.

AV. SIZE: 45–60 cm (tall)

HABITAT: All habitats across mainland Australia; absent in Tasmania

FIRST AID: Control any bleeding with pressure; clean area with antiseptic; may require antibiotics and/or tetanus immunisation; seek medical attention

D&D RATING: Treat with caution

Feral Dog *Canis lupus familiaris*

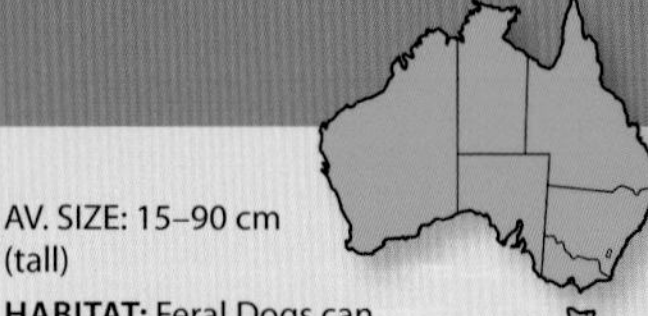

The loveable family dog is arguably Australia's most dangerous and deadly animal, with over 30,000 people seeking treatment for bites every year.

DESCRIPTION: There are over 800 recognised breeds. Compared to purebred Dingoes, dogs have proportionally smaller brains, skulls and teeth. Unlike Dingoes, dogs also have the ability to bark.

BEHAVIOUR: All dogs are highly social animals, although different breeds display variations in behaviour. Like the Dingo, the domestic dog is both a highly effective predator and scavenger.

DANGER TO HUMANS: Well-trained dogs pose little threat to humans, but abused animals (particularly aggressive breeds) can be dangerous. Abandoned dogs can become feral and a menace in outer suburban areas.

AV. SIZE: 15–90 cm (tall)

HABITAT: Feral Dogs can interbreed with Dingoes and occupy a similarly country-wide habitat, including fringe areas of towns/cities

FIRST AID: Control any bleeding with pressure; clean area with antiseptic; may require antibiotics and/or tetanus immunisation; seek medical attention

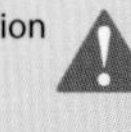

D&D RATING: Treat with caution

Feral Pig *Sus scrofa*

Introduced to Australia by early settlers, pigs soon escaped and became feral. In the ensuing years, they began causing large scale environmental problems.

DESCRIPTION: Feral Pigs are highly variable in colour and are coated in stiff, wiry bristles. They are generally leaner than domestic pigs. Large boars ("razorbacks") can weigh more than 100 kg.

BEHAVIOUR: Feral Pigs spend most of the day in the shelter of dense bush, emerging at night to forage. Both male and female pigs have stout protruding tusks, which are used to root for starchy tubers, fungi, worms and a variety of other foods.

DANGER TO HUMANS: Cornered or injured Feral Pigs can become extremely aggressive and a risk to hunters. They also carry many infectious diseases and internal/external parasites that can spread to humans.

AV. SIZE: 75-115 kg

HABITAT: Most habitats throughout Australia, except the arid zones

FIRST AID: Control any bleeding with pressure; clean area with antiseptic; may require antibiotics and/or tetanus immunisation

D&D RATING: Treat with caution

Water Buffalo *Bubalus bubalis*

Between 1825 and 1843, Asian Water Buffalo were imported into Australia's remote northern colonies as a source of meat. When these outposts were abandoned in the 1940s, buffalo numbers quickly increased.

DESCRIPTION: Water buffalo are thick, powerfully built bovines. Males (bulls) can weigh in excess of 600 kg. Two breeds exist in Australia — the "swamp" buffalo (with long, swept-back crescent horns) and the "river" buffalo (with curled horns).

BEHAVIOUR: In Asia, water buffalo have been successfully domesticated. They are usually placid but bulls may become agitated and unpredictable. They frequent waterholes, where they cause substantial environmental damage.

DANGER TO HUMANS: Water buffalos sometimes stray into the paths of fast-travelling cars, causing collisions that can be fatal to both buffalos and motorists. Water buffalos are also known carriers of tuberculosis and brucellosis, both of which pose a risk to Australia's commercial cattle industry.

AV. SIZE: 400–700 kg

HABITAT: Swamps, waterways and floodplains of Australia's tropical north

FIRST AID: If physical trauma results, control bleeding and seek medical attention as required

D&D RATING: Treat with caution

Birds

Not too many countries in the world are known for dangerous birds and Australia is no exception. In general, birds are not particularly dangerous animals. However, many species are territorial and they will attack human trespassers. Ultimately, this occurs during the breeding season when over-protective parents defend their chicks against potential threats.

The Australian Magpie (*Gymnorhina tibicen*) is found all around the country and is famous for its melodious carolling, backyard visits and territorial fury. During the nesting season — August to October — many local councils will display signs in parks and gardens warning people of imminent dive-bombing attacks from normally peaceful "maggies". Most attacks are used to intimidate intruders and, more often than not, no damage is done. A national survey on magpie attacks revealed that 90% of males and 72% of females surveyed have been attacked by a magpie at some point in their life.

Another bird notorious for antisocial behaviour during the breeding season is the Masked Lapwing (*Vanellus miles*). Like the magpie, this species is a common sight in suburban parks and gardens. Armed with a bright yellow spur on the carpal joint of each wing, this ground-dwelling species is highly alert and vigorously defends its chicks against dogs, cats, people and other birds (especially corvids).

Australia's most dangerous native bird is also one of the country's rarest. While the Southern Cassowary (*Casuarius casuarius*) has been known to attack visitors to its rainforest home, it seems unfair to brand such a magnificent — and endangered — species as dangerous. It faces constant threats from developers, motor vehicles, dogs, Feral Pigs (which compete for food and eat the cassowary's eggs), tropical cyclones and disease.

Top: School children are taught to watch for swooping Magpies — a national annoyance. **Right:** A Masked Lapwing's threat display.

Cassowary *Casuarius casuarius*

The flightless Southern Cassowary is Australia's heaviest bird and just one of three cassowary species found in the world. The cassowary is a powerful bird that has attacked humans. However, fewer than 1000 adult cassowaries survive in the wild and they cannot be considered a genuine danger to humans. The Mission Beach population suffered greatly from Cyclone Larry in 2006.

DESCRIPTION: Tall, heavy-set flightless bird with coarse, glossy black plumage, a bright blue neck and long, dangling red wattles. Together with its size and colouration, one of its most prominent features is a tall brown casque that crowns the head. The Southern Cassowary has a strong beak and three large toes on each of its thick, powerful legs. The middle toe is the longest (up to 12 cm), while the inside toe bears a long, sharp dagger-shaped claw.

BEHAVIOUR: Cassowaries are usually solitary birds that dwell in tropical rainforests, mangrove forests and melaleuca swamps of North Queensland. Each bird has a home range of up to 3 km^2 where it forages on fungi, fallen fruit, snails, insects and (sometimes) carrion. Their dung plays an important role in seed dispersal — they are one of the few fruit-eating animals to disperse large rainforest fruits.

DANGER TO HUMANS: Cassowaries don't take kindly to overly inquisitive humans and can become aggressive. These large birds pack a powerful kick that can cause serious injury to humans and their pets. In 1926 a teenage boy was killed by a cassowary after being kicked in the throat. This is the only human death on record.

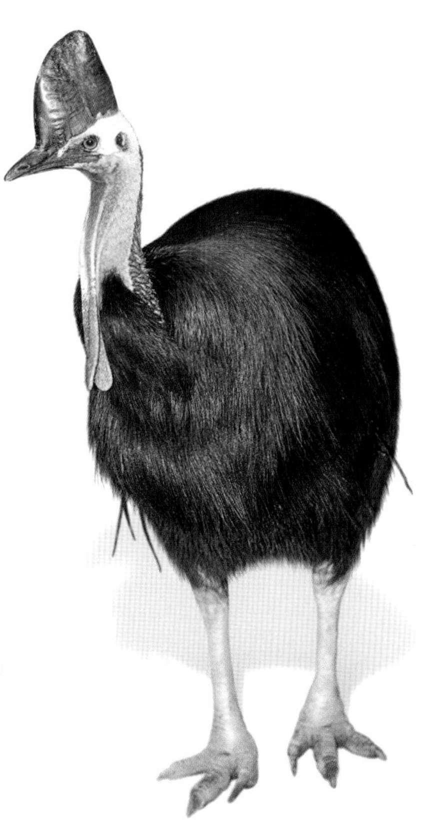

AV. SIZE: 1.75 m
HABITAT: Tropical rainforests

FIRST AID: Control any bleeding with pressure; clean area with antiseptic; seek medical attention; serious cases may require surgery
D&D RATING: Treat with caution

First Aid Information

While most people will avoid serious injury from Australia's animals, it is important to be well versed in first aid procedures, as it can make all the difference between life and death. Ideally, every home should stock a copy of *Australian First Aid*, which is authorised by St John Ambulance and specifically designed for Australian conditions. Listed below are some of the key first aid procedures for encounters with Australia's potentially dangerous and deadly wildlife.

"DRABCD"

DRABCD is always your initial action plan. It represents six procedures (Danger, Response, Airway, Breathing, Circulation and Defibrillation) that you must immediately follow in any accident or emergency situation.

DANGER — Check for Danger

- Check for danger to yourself, the people around you and the casualty.
- Always make sure there is no danger to yourself or others before proceeding — you don't want to become a casualty yourself or have to deal with any extra unnecessary casualties.
- Only proceed if it is safe to do so.

RESPONSE — Check for a Response

- Check for a response. Is the casualty conscious?
- Gently squeeze the casualty's hand or shoulders and talk loudly to see if he or she is conscious. Ask the casualty to squeeze your hand.
- If the casualty is conscious, check and manage specific injuries accordingly (e.g. snake bite).
- If the casualty is unconscious, place them on their side in the recovery position and ask a bystander to phone triple zero (000) or send them for help.

AIRWAY — Clear and Open the Airway

- With the casualty in the recovery position, tilt the head backwards and slightly down to drain any fluids.
- Clear any foreign objects from the casualty's mouth with your fingers. (Note: only remove dentures if loose or broken.)

BREATHING — Check for Breathing

- Look for rising and falling chest movements.
- Place your head close to the casualty's mouth. Listen for breathing sounds and feel for breath on your cheek.
- If the casualty is breathing, check and manage specific injuries accordingly (e.g. snake bite).
- If there are no signs of life and the casualty is unconscious, not breathing and not moving, start CPR by giving two initial breaths.

CPR — (Cardiopulmonary Resuscitation)

- CPR involves giving 30 compressions at a rate of approximately 100 compressions per minute followed by two breaths.

DEFIBRILLATE

- Apply a defibrillator (if available). A defribrillator is a medical machine that applies electric shock to the heart.
- Follow the voice prompts it provides. Or continue CPR until medical assistance arrives.

Pressure Immobilisation Bandage

A pressure immobilisation bandage is the recommended treatment for envenomation by snakes, cone shells, blue-ringed octopuses, funnelweb spiders and mouse spiders. It is also recommended for bee, wasp and ant stings in allergic individuals. This first aid method was developed in 1975 at the Commonwealth Serum Laboratories (CSL Ltd) with the purpose of "containment" — that is, stopping venom introduced at the bitten area from spreading to other parts of the body. The following procedure is recommended:

- Phone triple zero (000) or send for medical assistance.
- Keep the victim and the bitten limb as still as possible. (Note: Do not remove any clothes as any movement may help circulate the venom.)
- Firmly apply a broad crepe bandage to the bitten area as soon as possible. If you do not have a proper bandage, cut any clothing, towels or similar items into strips and use these.
- Bandage upwards from the lower portion of the bitten area.
- Extend the bandage as far up the limb as possible.
- Apply a splint to the limb and bind it firmly with a bandage. If you do not have a proper splint, use any rigid object (e.g. tree branch, axe handle, rolled up newspaper).

The following points should always be remembered:

- **Do not use a pressure immobilisation bandage to treat envenomation by stinging fish, scorpions, centipedes, sea jellies or any spiders other than funnelwebs and mouse spiders.**
- **Do not suck, cut or wash venom from the bitten area. Any residual venom will help medical professionals identify the animal responsible and administer the correct antivenom.**
- **Do not try to catch or kill the animal.**
- **Do not apply the bandage so tightly that it cuts off circulation.**
- **Do not use a constrictive tourniquet.**

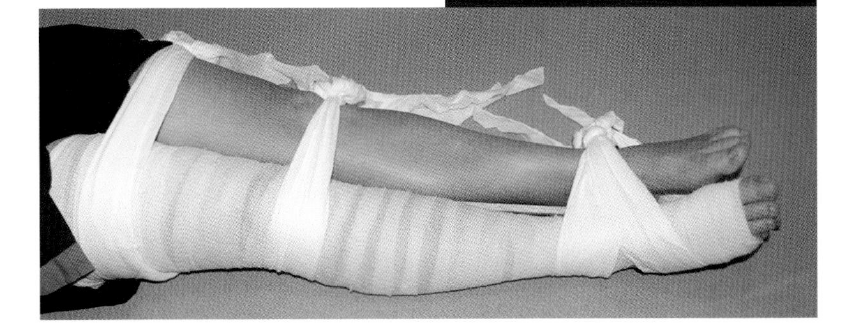

Severe Bleeding

A shark, crocodile, canine or moray eel bite may cause severe physical trauma and bleeding. This can result in a panic-stricken patient and a medical emergency that must be dealt with as quickly and calmly as possible. Often the stress involved in such extreme situations can have a strong emotional and psychological impact, which may require later treatment of its own. In cases of severe bleeding it is vital to phone an ambulance or send for medical assistance and apply direct pressure to the wound. The following first aid procedure is recommended:

- Lay the casualty down. Handle gently if you suspect any fractures.
- Elevate the injured area above heart level.
- Using your whole hand, apply direct pressure to the wound. (Note: If possible, use surgical gloves and place sterile padding or clean clothing over the wound.)
- If possible, squeeze the edges of the wound together. Apply direct pressure for at least ten minutes without checking to see if the bleeding has stopped.
- Secure the wound with a bandage. (Note: Make sure any padding remains in place.)
- If bleeding continues, leave initial padding in place, apply more padding and secure this with a bandage.
- If bleeding continues, replace second padding and bandage.
- Treat the casualty for shock.

Shock

Physical shock is a serious medical condition that occurs when an insufficient amount of blood circulates through the body. Physical shock is therefore different from mental shock, although both states can occur in an emergency. It is important to recognise the symptoms of hypovolemic shock (the most common type of shock) which include a weak and rapid pulse, cold and clammy skin, rapid and shallow breathing, faintness, dizziness, nausea and pale skin. The following first aid procedure is recommended:

- Follow procedures for DRABCD.
- Phone triple zero (000) or send for medical assistance.
- Rest the casualty in a comfortable position and elevate legs above heart level. (Note: Do not raise legs if they are fractured or casualty has suffered a snake bite).
- Loosen tight clothing around the neck, chest and waist.
- Keep the casualty warm.
- Treat other injuries accordingly.
- Monitor and record pulse and breathing.

Right: Prevention is better than cure. Always wear boots and long trousers when in potential snake areas. **Below:** It is important to have a well equipped first aid kit stocked with a range of bandages.

Glossary

ANALGESIC A remedy that relieves or removes pain.

ANTIVENOM (Also called "antivenin" or "antivenene"). An antitoxin, made from the antibodies produced naturally by an animal such as a horse, that is active against the venom of a spider or snake or other venomous animal.

ARANEOMORPH One of two infraorders of spiders. Araneomorphs are the more modern of the two.

ATRAXOTOXIN The toxin produced by the funnelweb spider.

CIGUATERA Food poisoning caused by eating fish or seafood contaminated with ciguatoxin.

CIGUATOXIN The general poisonous effect of toxic single-celled marine organisms called dinoflagellates when eaten by an animal. When eaten by one fish, then another, the toxin concentrates in the largest fish. If humans eat these big fish, they suffer gastrointestinal and neurological symptoms.

DIURNAL Active during the day.

ENDEMIC Native to a country or locality.

ENVENOMATION Poisoning that results from a bite or sting (i.e. an injection of venom) from an animal.

GASTROPODS Any of the Gastropoda, a class of molluscs comprising the snails, having a shell of a single valve, usually spirally coiled, and a ventral muscular foot on which they glide about.

MANDIBLES One of the first pair of mouth-part appendages, typically a jaw-like biting organ, but styliform or setiform in piercing and sucking species.

MYGALOMORPH One of two infraorders of spiders. Mygalomorphs are the more ancient of the two.

PELAGIC Living at or near the surface of the ocean, far from land, as certain animals or plants.

POISON A substance, which is toxic when breathed in or eaten.

POLYPS A sedentary type of animal form characterised by a more or less fixed base, columnar body, and free end with mouth and tentacles, especially as applied to coelenterates.

PROBOSCIS An elongate but not rigid feeding organ of certain insects formed of the mouthparts, as in the Lepidoptera and Diptera.

SPICULE A small or minute, slender, sharp-pointed body or part.

STINGER The colloquial name for a sea jelly with a serious sting.

TERRESTRIAL Living on the ground; not aquatic, arboreal, or aerial.

VENOM A substance that is toxic when it makes contact with the tissue underneath the skin, for example, after skin puncture by fangs. Venom is generally harmless if eaten.

VIPER Any snake of the family Viperidae. These snakes occur throughout Africa, Europe, Asia and the Americas.

ZOOIDS Any animal organism or individual capable of separate existence, and produced by fission, gemination or some method other than direct sexual reproduction.

Below: The king of stealth — the Estuarine Crocodile.

Index

Index continued

Links & Further Reading

Books

Brunet, B. *Spider Watch: A Guide to Australian Spiders*, Reed New Holland, Sydney, Australia, 1996

Czechura, G. *Amazing Facts about Australia's Deadly and Dangerous Wildlife*, Steve Parish Publishing, Brisbane, Australia, 2007

Daley, RK, Stevens, JD, Last PR & Yearsley GK. *Field Guide to Australian Sharks and Rays*, CSIRO Publishing, Canberra, 2002

Edmonds, C. *Dangerous Marine Creatures*, Reed Books, Frenchs Forest, New South Wales, 1989

Sutherland, S & Tibballs, J. *Australian Animal Toxins (2nd edition)*, Oxford University Press, Melbourne Australia, 2001

Sutherland, S & J. *Venomous Creatures of Australia (5th edition)*, Oxford University Press, Melbourne, Australia, 2006

Swanson, S. *Field Guide to Australian Reptiles*, Steve Parish Publishing, Brisbane, Australia, 2006

White, J, Edmonds, C & Zborowski. *Australia's Most Dangerous Spiders, Snakes and Marine Creatures*. Australian Geographic Pty Ltd, Terrey Hills, New South Wales, 1998

Websites

IAMS — Dangerous Marine Animals of Northern Australia **www.aims.gov.au/pages/research/project-net/dma/pages/dma-01.html**

Australia Zoo **www.australiazoo.com.au/**

Australian CSL Ltd Antivenom Handbook — Online Edition **www.toxinology.com/generic_static_files/cslavh_contents.html**

Australian Museum Online — Fact Sheets **www.amonline.net.au/factsheets/**

Australian Reptile Park **www.reptilepark.com.au/**

Australian Museum Online — Fishes **www.amonline.net.au/fishes/index.cfm**

Australian Venom Research Unit **www.avru.org/**

CSIRO — Australian National Insect Collection Fact Sheets **www.csiro.au/csiro/channel/pchey.html**

International Shark Attack File **www.flmnh.ufl.edu/fish/sharks/isaf/isafabout.htm**

Museum Victoria — Spiders **www.museumvictoria.com.au/Spiders/**

Queensland Museum — Dangerous Snakes **www.qm.qld.gov.au/features/snakes/dangerous/index.asp**

St John Ambulance Australia — Quick Reference Guide **www.stjohn.org.au/**

Acknowledgments

St John First Aid Protocols are for the Australian market only. All care has been taken in preparing the information but St John takes no responsibility for its use by other parties or individuals. The information is not a substitute for first aid training. For more information on St John first aid training and kits visit **www.stjohn.org.au** or call **1300 360 455**

Opposite: The beauty of the blue-ringed octopus belies its danger.

Published by Steve Parish Publishing Pty Ltd
PO Box 1058, Archerfield, Queensland 4108
Australia

www.steveparish.com.au

ISBN 978174193405 2

First published 2008

Principal Photographer: Steve Parish

Additional photography: Gilson Francois/Bios-Auscape: p. 79 (bottom); Tony Ayling: p. 31 (top); Michael Cermak: pp. 11, 65, 67 & 77 (top); CDC/PHIL/Corbis: p. 79 (top); DK Limited/Corbis: p. 76 (bottom); Ian Banks/Diving the Gold Coast: p. 35 (bottom); Greg Harm: p. 74 (top); Patrick Honan: pp. 3, 6, 66 & 76 (top); IStockPhoto: p. 7 (bottom right); David Muirhead/Marine Themes: p. 57 (bottom); M & I Morcombe: p. 86 (bottom); Ian Morris: pp. 2, 5 (top right & bottom inset), 12 (bottom), 13, 18, 24, 25 (bottom), 27, 70 (top), 71, 80 & 91 (right); Newsource: p. 5 (top left); Gary Bell/OceanwideImages.com: p. 52; Qld Museum: pp. 19 (top), 70 (bottom) & 73; Bruce Cowell/Qld Museum: p. 38 (top); Jeff Wright/Qld Museum: pp. 4, 10, 16 (top), 34, 64, 72, 75 (bottom), 78 & 81; Gunther Schmida: p. 57 (top); Dr. Jamie Seymour: p. 33; Peter Slater: p. 77 (bottom); Gary Steer: p. 84 (bottom); Steve Swanson: pp. 17 (bottom), 19 (bottom), 21 (top) & 23 (top); Ron & Valerie Taylor: pp. 5 (centre), 8-9, 26, 28 (bottom), 29-30, 31 (centre right), 32, 41 (top), 42 (top), 43-47, 48 (top & bottom left), 49, 50-51, 53, 58 (bottom), 62 (top right & centre right); Clare Thomson: pp. 7 (top left-right), 89 & 90-91

Front cover image: White Shark, Ron & Valerie Taylor

Title page: Grey Reef Sharks, Ron & Valerie Taylor. Inset, top to bottom: Red-bellied Black Snake, Ian Morris; Wolf Spider, Jeff Wright/Queensland Museum

Text: Ted Lewis
Editorial: Sarah Lowe; Kerry McDuling; Michele Perry & Helen Anderson, SPP
Design: Gill Stack, SPP
Image Library: Clare Thomson, SPP
Production: Tina Brewster, SPP

Prepress by Colour Chiefs Digital Imaging, Brisbane, Australia
Printed in Singapore by Imago

Produced in Australia at the Steve Parish Publishing Studios